Time for mapping

Manchester University Press

# Time for mapping

## Cartographic temporalities

EDITED BY SYBILLE LAMMES, CHRIS PERKINS, ALEX GEKKER, SAM HIND, CLANCY WILMOTT AND DANIEL EVANS

Manchester University Press

Copyright © Manchester University Press 2018

While copyright in the volume as a whole is vested in Manchester University Press, copyright in individual chapters belongs to their respective authors.

An electronic version of this book is also available under a Creative Commons (CC-BY-NC-ND) licence, thanks to the support of EU Horizon 20/20 project, which permits non-commercial use, distribution and reproduction provided the editor(s), chapter author(s) and Manchester University Press are fully cited and no modifications or adaptations are made. Details of the licence can be viewed at https://creativecommons.org/licenses/by-nc-nd/4.0/.

Published by Manchester University Press
Altrincham Street, Manchester M1 7JA

www.manchesteruniversitypress.co.uk

British Library Cataloguing-in-Publication Data
A catalogue record for this book is available from the British Library

ISBN  978 1 5261 2253 7  hardback
ISBN  978 1 5261 2252 0  open access

First published 2018

The publisher has no responsibility for the persistence or accuracy of URLs for any external or third-party internet websites referred to in this book, and does not guarantee that any content on such websites is, or will remain, accurate or appropriate.

Typeset by
Servis Filmsetting Ltd, Stockport, Cheshire
Printed in Great Britain by
Lightning Source

# Contents

| | |
|---|---|
| *List of figures* | vii |
| *List of tables* | x |
| *Notes on contributors* | xi |
| *Acknowledgements* | xiv |

1  Introduction: mapping times  1
   *Alex Gekker, Sam Hind, Sybille Lammes, Chris Perkins and Clancy Wilmott*

## Part I: Ephemerality/mobility  25

2  Nodes, ways and relations  27
   *Joe Gerlach*

3  Mapping the quixotic volatility of smellscapes: a trialogue  50
   *Sybille Lammes, Kate McLean and Chris Perkins*

4  Seasons change, so do we: heterogeneous temporalities, algorithmic frames and subjective time in geomedia  91
   *Pablo Abend*

## Part II: Stitching memories  113

5  'Space-crossed time': digital photography and cartography in Wolfgang Weileder's *Atlas*  115
   *Rachel Wells*

6  Traces, tiles and fleeting moments: art and the temporalities of geomedia  138
   *Gavin MacDonald*

7  Digital maps and anchored time: the case for practice theory  154
   *Matthew Hanchard*

## Part III: (In)formalising  173

8  Mapping the space of flows: considerations and consequences  175
   *Thomas Sutherland*

| 9  | Maps as foams and the rheology of digital spatial media: a conceptual framework for considering mapping projects as they change over time | 197 |
|---|---|---|
|   | *Cate Turk* |   |
| 10 | Maps as objects | 223 |
|   | *Tuur Driesser* |   |
| 11 | From real-time city to asynchronicity: exploring the real-time smart city dashboard | 238 |
|   | *Michiel de Lange* |   |
| 12 | Conclusion: back to the future | 256 |
|   | *Alex Gekker, Sam Hind, Sybille Lammes, Chris Perkins and Clancy Wilmott* |   |

# Figures

| | | |
|---|---|---|
| 1.1 | 'Spatial Turn' over time (Google Books Ngram Viewer, http://books.google.com/ngrams) | 9 |
| 2.1 | GPS route tracing around Witham, Essex | 31 |
| 2.2 | Fieldwork scrapbook extract from author participation in OSM mapping party | 33 |
| 2.3 | GPS route tracing author's movement south-easterly while mapping Witham | 33 |
| 2.4 | Author's train journey photograph demonstrating the train operator's use of OSM base maps in passenger navigation screens | 34 |
| 2.5 | Author GPS track | 35 |
| 2.6 | Author GPS track | 40 |
| 2.7 | Author GPS track | 44 |
| 3.1 | Smellmap: Edinburgh | 51 |
| 3.2 | Smellcolour sketch: Brooklyn | 56 |
| 3.3 | Smellmap: Pamplona (still image from movie) | 57 |
| 3.4 | Smellmap Newport, Rhode Island | 65 |
| 3.5 | Smellscaper App Smell Notes interface | 68 |
| 3.6 | Spanish language smell notes prior to classification | 73 |
| 3.7 | Contrasting smellscapes of the West End of London, mashup showing emissions, nature, food and animal smells against an OSM backdrop, as recorded in social media | 76 |
| 4.1 | Compilation of stills from the movie *Crank* (dir. Mark Neveldine and Brian Taylor/Lakeshore Entertainment, Lions Gate Films, RadicalMedia, GreeneStreet Films/USA/2006) | 98 |
| 4.2 | Seasons change by half-turns in Brooklyn, New York. Google Earth images | 105 |
| 5.1 | Wolfgang Weileder, *Gulf of Naples s2912*, 2009, from the *Seascapes* series. Lambda Print, 75 x 154cm | 116 |
| 5.2 | Hiroshi Sugimoto, *Seascape: North Atlantic Ocean, Cape Breton*, 1996 | 118 |
| 5.3 | Hiroshi Sugimoto, *Henry VIII*, 1999, from the *Portraits* series. Gelatine silver print, 149.2 x 119.4cm | 119 |

| | | |
|---|---|---|
| 5.4 | Wolfgang Weileder, *Piazza San Pietro, Rome, Slice 1756*, 2010, from the *Atlas* series, 2011–present. Archival inkjet print, 140 x 22cm | 120 |
| 5.5 | Wolfgang Weileder, *Atlas*, exhibition view, Northern Gallery for Contemporary Art, 2013 | 121 |
| 5.6 | Wolfgang Weileder, *House-Madrid*, 2004. Gelatine silver print, 79 x 100cm | 122 |
| 5.7 | Screen grab from Google Maps, December 2014 | 126 |
| 5.8 | Wolfgang Weileder, *Place des Vosges, Paris, Slice 2356*, 2012, from the *Atlas* series, 2011–present. Archival inkjet print, 137 x 233cm | 127 |
| 5.9 | Jon Rafman, *D52, Blaru, France*, 2011 from The Nine Eyes of Google Street View | 128 |
| 5.10 | Wolfgang Weileder, *Skydeck*, 2009, installation in the Workplace Gallery, Gateshead | 130 |
| 5.11 | Wolfgang Weileder, *La Terme, Piazza Oberdan, Milan*, 2008 | 131 |
| 5.12 | Wolfgang Weileder, *Res Publica*, 2012, 5x5 Public Art Festival, Washington DC | 132 |
| 5.13 | Wolfgang Weileder, *Res Publica*, 2012, 5x5 Public Art Festival, Washington DC | 133 |
| 5.14 | Wolfgang Weileder, *Camera*, 2002 | 134 |
| 5.15 | Wolfgang Weileder, *Camera*, 2002. Fabric, aluminium, optics, 220 x 780 x 420cm | 134 |
| 6.1 | *Amsterdam RealTime* (2002) Esther Polak, Jeroen Kee and Waag Society. Participatory mapping project and installation; this image depicts a composite of GPS traces produced by participants | 140 |
| 6.2 | Detail from *Monochrome Landscapes* (2004) Laura Kurgan. Yellow: southern desert, south-eastern Iraq, between Al Busayyah and An Nasiriyah | 147 |
| 6.3 | Coronado Feeders, Dalhart, Texas, from *Feedlots* (2013) Mishka Henner | 147 |
| 8.1 | Unknown author, *Tabula Peutingeriana*, c. fourth–fifth centuries | 177 |
| 8.2 | J. C. R. Colomb, map of the British Empire from 1886. Norman B. Leventhal Map Center Collection | 178 |
| 8.3 | Charles Minard, Carte Figurative, 1869 | 180 |
| 8.4 | Gerardus Mercator, world map, 1569 | 187 |
| 9.1 | Wind map, showing the way winds are flowing around the Earth (https://earth.nullschool.net) | 199 |
| 9.2 | Haiyan/Yolanda Swipe Map, enabling comparison of before and after satellite imagery (www.esri.com/services/disaster-response/hurricanes/typhoon-haiyan-yolanda-swipe-map) | 199 |

| | | |
|---|---|---|
| 9.3 | Differentiating relations in a network (from J. C. Plantin, 'The Fukushima online issues map: linking practices between spheres of actors', *Cartonomics: Space, Web and Society*, 2012) | 203 |
| 9.4 | Assemblages – from network to foam. Based on original map by S. Adler, G. Glasze, C. Bittner and C. Turk | 205 |
| 9.5 | The HOT mapping deployment post Typhoon Haiyan/Yolanda: inception and growth | 213 |
| 9.6 | The HOT mapping deployment post Typhoon Haiyan/Yolanda: discussion and disruption | 213 |
| 9.7 | The HOT mapping deployment post Typhoon Haiyan/Yolanda: viewed through other actors | 214 |
| 9.8 | Mapping software collating crowd-sourced reports about storm damage (Volontaires internationaux en soutien aux opérations virtuelles, https://haiyan.crowdmap.com) | 215 |
| 9.9 | Foam interfaces can be thick with multiple temporalities | 216 |
| 9.10 | Map project histories, with apologies to Hokusai | 217 |
| 9.11 | Crowd-sourced reporting of water heights. Information is current for a limited time. Philippine Flood Map 2013, https://philfloodmap.crowdmap.com | 218 |

# Tables

| | | |
|---|---|---|
| **3.1** | Published 'Smellmaps' by Kate McLean | 52 |
| **9.1** | Archival mapping sources used by author for analysis | 209 |
| **9.2** | Questioning the mapping bubbles and form | 210 |

# Contributors

**Pablo Abend** is Scientific Coordinator of the interdisciplinary research school 'Locating Media' at the University of Siegen, Germany. His research interests are geomedia, localised and situated media research, game studies, qualitative methods in media studies, and science and technology studies. He is co-editor of the journal *Digital Culture & Society* http://digicults.org/.

**Michiel de Lange** is Assistant Professor in New Media Studies, Department of Media and Culture Studies, Utrecht University, Netherlands; co-founder of The Mobile City (www.themobilecity.nl), a platform for the study of new media and urbanism; and works as a researcher in the field of (mobile) media, urban culture, identity and play.

**Tuur Driesser** is a PhD candidate at the Centre for Interdisciplinary Methodologies, University of Warwick. His research examines how smart cities are made through digital maps and data visualisations, exploring different methods and methodologies for the study of, with and through maps.

**Daniel Evans** is a PhD student in Human Geography at the University of Manchester. His doctoral research explores social inequalities and race in the context of white masculinities in the UK and he maintains a keen interest in cartography.

**Alex Gekker** is Lecturer in the departments of Media and Culture at the University of Amsterdam and Media & Communication at Erasmus University Rotterdam. He completed his PhD at Utrecht University, working on the relations between mapping, digital interfaces and power, and is interested in ways socio-technical systems are designed to influence users, with research also addressing quantification and datafication of society, the experience economy and interface critique.

**Joe Gerlach** is Lecturer in Human Geography at the School of Geographical Sciences, University of Bristol. His research interests span cultural and political geography, including critical cartography, micropolitics, non-representational theory and nature–society relations in Ecuador.

**Matthew Hanchard** is a PhD student at Newcastle University (Geography, Politics and Sociology), developing a practice orientated digital sociology of maps. He explores mundane digital map engagements and the extent to which they anchor other mobilities.

**Sam Hind** is Research Assistant at the University of Siegen. He is a member of the Playful Mapping Collective, and co-author of *Playful Mapping in the Digital Age* (Institute of Network Cultures, 2016). He is interested in the risks, politics and futures of digital navigation, as well as the pedagogic value of playful methodologies.

**Sybille Lammes** is Professor of New Media and Digital Culture at Leiden University. Her background is in media studies, which she has always approached from an interdisciplinary angle, including cultural studies, science and technology studies, and critical geography.

**Gavin MacDonald** is Senior Lecturer on the Undergraduate Programmes in Art Theory and Practice at Manchester School of Art. His research deals with the art and visual culture of geomedia.

**Kate McLean** is Programme Director for Graphic Design at Canterbury Christ Church University, UK. She is a designer, mapper and collector of urban smells working with visual and olfactory communication media.

**Chris Perkins** is Reader in Geography at the University of Manchester. He is interested in how people deploy mapping and in particular researches ethnographies of the map, with an emphasis upon playful and critical approaches to different everyday contexts.

**Thomas Sutherland** is Lecturer in Media Studies at the University of Lincoln. His research focuses primarily on the interstices between continental philosophy and media theory, with a particular interest in concepts of mediation, representation, communicability and time, as well as political normativity and media history.

**Cate Turk** is completing a PhD in Geography at the Friedrich Alexander University of Erlangen-Nuremberg, and is currently Project Officer for the

'Rivers of Emotion' project at the ARC Centre of Excellence for the History of Emotions, at the University of Western Australia. Her work examines various aspects of contemporary cartographic practice including collaboration, temporality and emotion.

**Rachel Wells** is Lecturer in Art History and Theory at Newcastle University, UK. Her research on issues of scale and distance in contemporary globalised art has been published widely, including her 2013 book *Scale in Contemporary Sculpture* (Routledge), and the 2017 publication *In Focus:* Static *2009 by Steve McQueen* (Tate).

**Clancy Wilmott** is Lecturer in Geography at the University of Manchester, and recently completed her PhD on the indeterminacies of mobile mapping in Sydney and Hong Kong. Her research centres on practices and disruptions in the flow of cartographic reason.

# Acknowledgements

We would like to express our gratitude to all those involved in the making of this book, from the individual authors who contributed their works and ideas, to those who hosted sessions and provided criticisms to help strengthen this publication. In particular, we would like to thank Nanna Verhoeff for her ideas in initially developing this project, as well as Joe Gerlach for hosting the workshop on which the book was largely based. We would also like to acknowledge Erasmus+ for contributing to the development of chapter 3, and thank Kate McLean for facilitating an artistic interpretation of 'smell map' cartography. Kate also assisted greatly in producing a novel layout of the trialogue discussion. The quality of the ideas expressed by authors has benefitted hugely from the critical input of Dan Evans.

The research leading to these results has received funding from the European Research Council under the European Community's Seventh Framework Programme (FP7/2007–2013) / ERC Grant agreement n° 283464. The results have also been partly funded by the European Commission under the Erasmus+ Key Action 2 Strategic Partnership funding framework, grant number 2014–1–UK01–KA203–001642.

# Abbreviations

| | |
|---|---|
| ANT | actor-network theory |
| API | Application Programming Interface |
| GIS | Geographical Information Systems |
| GPS | Global Positioning System |
| HOT | Humanitarian OpenStreetMap Team |
| NGO | non-governmental organisation |
| OSM | OpenStreetMap |

# 1

## Introduction: mapping times

*Alex Gekker, Sam Hind, Sybille Lammes, Chris Perkins and Clancy Wilmott*

Digital mapping, though generally conceived as a spatial activity, is just as strongly grounded in time. The digital era has disintegrated the representational fixity of maps, and instead given rise to maps that shift with each moment and movement. Scholars, adept at grappling with the spatial implications of digitality, continue to struggle to conceptualise and communicate the temporal consequences of maps. In this collection, we seek to take up Doreen Massey's (2005: 107) still critical concern: how do we cope with maps as mediators of the 'ongoing stories' in the world? Mapping has long wrestled with enrolling time into such narratives. This collection examines how this difficulty is impacted by the presence of digital mapping technologies that, arguably, have disrupted our understanding of time as much as they have provided coherence. The contributions in this book move beyond the descriptive to pay particular attention to what might be called the 'critical dynamics' of time.

We, and other authors in this book, suggest that the relation between digital mapping and its temporalities should be conceived as plural, dynamic and situated. Also, as digital mappings are approached in this book from an interdisciplinary angle – as medial, cartographic and technological practices – different scholarly perspectives reveal different understandings of temporalities. These twin concerns with dynamism, and plural responses to dynamism, are the central foci of this volume. The chapters in this book reflect this multiplicity of tempo-spatialities rather than spatio-temporalities.[1] Many of the chapters implicitly reflect on Merriman's (2011) challenge that notions of 'space-time' and 'time-space' have frequently rested on rather static conceptions. In proposing 'movement-space' as an alternative, however, Merriman remains inattentive to the *ontological* proclivities of particular *digital* formations. Although he does not invoke digital mapping in his re-theorisation, and although few of our authors directly cite his work, the implications of his intervention are pertinent issues consistently and carefully addressed in this book, precisely because the digital reveals the contingency, mutability and dynamism of mapping practices.

Our authors have diverse scholarly backgrounds and use these to investigate different cultural mapping practices in terms of ephemeralities, 'time's arrow', (a)synchronicities, rhythms and velocities. Together, the chapters offer a broad spectrum of methodologies and conceptual frameworks to help us understand the rich texture of relations between digital mapping and temporalities. This book also proposes that digital maps bring new temporal affordances into play for users through their cartographic interfaces. It is through these embedded and interactive affordances that digital mapping transforms our notions of immediacy and futurity; allowing us to track our current and past locations as well as calling the future into being by advising on potential routes and ways forward.

## Introducing the future?

The 21 October 2015 was *Back to the Future* Day – the destination date referenced in Robert Zemeckis' iconic 1980s films, punched into the DeLorean time machine as the characters travel 'back to the future'. The twenty-six year gap between the November 1989 release of the film, and the imaginary Hill Valley[2] has now disappeared into the conundrum of history. However, at the time, this imagined future was safely enough removed from the very different 'home ages' of the film, allowing entertaining paradoxical contrasts between the past (1885 and 1950s America), the present (1985) and the future. This vision of the (now passed) future in 2015 acts as a temporal map of sorts; a marker in the various possible timelines yet to manifest themselves.

The films speak to our collection of chapters, not only because of this mutability and its relation to the imagined temporalities and spatialities of Hill Valley, but also because of their status as cultural markers relating to North American small-town life in the post-war era from the vantage point of 1985, and to the ways we at once remember the past, but also re-appropriate it to change the future (Ní Fhlainn, 2010). It plays with central tropes associated with time: 'time's arrow' is subverted as Doc and Marty are transported back, and to the future, to change events; the episodic structure of the trilogy of films and their duration reminds us that time plays out, but is also commodified, as a cultural and embodied experience screening in a multiplex, evoking particular emotions.

In light of this volume, *Back to the Future* also draws attention to important aspects of mapping as a device to carry people back into the past, and forward into the future. The technologies deployed in the film, such as the hoverboard and the DeLorean time machine, carry protagonists into possible futures in the same way that mapping transports people who deploy it, bringing new events to life. This kind of technologically driven future is analogous to recent Marxist and accelerationist literature concerned with futurity, for example Mason (2015) and Srnicek and Williams (2015). The plot devices in the film playfully draw together the synchronic and asynchronic, the past, the present and the future, and show how rhythms of everyday mapping encounters have consequences; the tracks of Marty interact with other characters, but sometimes not at the same time. The contingent nature of time itself becomes the storyline and also does its work through a particular temporality. This temporality invites the viewer to question links between the real and the 'reel'. Other cultural forms evoke the film(s), as: spin-off series, computer games and websites through which fans relive their encounters with the original. Websites map the locations in Los Angeles where scenes were shot, highlighting the practical linking of the imagined and the real through a mashup;[3] a digital mapping form that serves as a way into understanding the placed temporalities of past, present and future, which supports so called 'set-jetting' practice (Joliveau, 2009).

By the 1980s, digital mapping had slowly begun to supplant the dominance of printed maps, albeit in ways that were much less ephemeral than contemporary applications: technologies were deployed on desktops to fix and freeze possible futures, a means to an instrumental end. Affordances remained static – paper maps were increasingly made by deploying digital technologies, with users discernible from producers, in space and time. Maps were mobile things that could be deployed in different places, but the people deploying them were separate from the display – acting *after* the event – prompted by, yet not inhabiting an ever-changing map.

2015 was also the ten-year anniversary of the launch of Google Maps; a seductive and intuitive interface, accessible from touch-screens over most of this period. This has allowed users to navigate and produce the world in novel ways, promising a pervasive and always-on control over space. The 'slippy map' seamlessly facilitated a dynamic interaction with the interface and, with the launch of the first iPhone in 2007, increasingly came to be deployed on smartphones; carrying the map user with them and bringing into being new mobile possibilities. Throughout that period, Google Maps rose to become the *de facto* mapping platform for millions around the world. Fighting off direct challenges from rival technology companies such as Yahoo, Microsoft and Apple (as well as from several regulators and government agencies), Google also contested interventions from the likes of OpenStreetMap (OSM), and acquired an ever-growing number of technology start-ups from Waze to Skybox Imaging, to strengthen its position as the most powerful digital mapping platform. As part of its ever-mutating strategy, however, Google sprung a surprise in 2015: quietly choosing to release its previously $400 per year Google Earth Pro software for free. The extent of the differences between Google Earth Pro and the free version were somewhat minimal, with Pro users able to import, print, capture and measure more extensively than their 'basic' counterparts. Clearly, our expectations for what a digital mapping service can and should allow us to do, has dramatically changed. Where once it required a $400 annual subscription to import up to 2,500 addresses and overlay historical traffic data, it now only needs a download installer and a licence key.

While these differences are superficially ones of cost, access and availability – between $400 and nothing at all – they are also indicative of *temporal* shifts in data acquisition, download speed, bandwidth capacity and user experience, as computer processing power, memory and storage have dramatically improved. Arguably, then, we are moving into another distinctive phase of digital life, characterised by ever-more novel functions designed to cater to temporal desires: 'slidable' Google Street View images, 'dynamic' in-train maps and 'rapid response' crisis management tools. These digital mapping transformations are no longer merely possible, they are to be *expected*. Far from the science fiction of *Back to the Future*, they are mere technological facts. Some have tentatively called this emergent landscape the 'post-digital' (Andersen, Cox and Papadopolous, 2014; Berry and Dieter, 2015), not only to account for the truism that digital software, platforms, apps and services comprise 'the everyday', and are now 'everywhere' and 'every*ware*' (Greenfield, 2006), but that there is now little 'unique' about this digital imbrication. As Cramer (2014: no pagination) suggests:

> pragmatically, the term 'post-digital' can be used to describe either a contemporary *disenchantment* with digital information systems and media gadgets, or a period in which our *fascination* with these systems and gadgets has become *historical* just like the dot-com age ultimately became historical in the 2013 novels of Thomas Pynchon and Dave Eggers. (emphasis added)

While these kinds of proclamations are certainly far from the whole truth – digital divides continue to abound – they nonetheless point towards a consolidation of now ten-year old digital platforms, and the generation of an *even newer* 'new normal' (Bratich, 2006: 493). In this changing context, the types of digital mapping platforms noted above, and talked about in much more detail within this volume, have brought to bear a whole plethora of new temporal phenomena that situate and populate, activate and agitate, and even enervate and incapacitate the contemporary world.

Together, these two cultural references, the imagined temporalities of *Back to the Future* and the mutable affordances enacted by the Google mapping platform, signal an important need for re-engaging with temporality – for at once going back to the future, but also for a slippy re-mapping of temporality. That is the project this volume addresses – but first what exactly might temporality entail in this context, and how has it been imagined?

## Thinking the temporal

Barbara Adam (2008) highlights seven key aspects of time, which, together, encompass important ideas implicit throughout this volume. Only chapter 11 explicitly adopts Adam's typology but themes charted by authors in our collection pick up on many of her concerns. She focuses on time frame, temporality, timing, tempo, duration, sequence and modalities. *Time frame* concerns the scale of analysis: dividing time up into units of varying scales, such as a day or a geological era. *Temporality* is much more procedural, and invites a focus on mutable qualities that unfold in human experience. *Timing* concerns the particularity of time: it emphasises events and moments instead of processes. *Tempo* is about speed: how fast something happens. *Duration*, on the other hand, concerns how long something lasts, whereas *sequence* addresses order and priority of events. *Modalities* considers the past, present and future in which time plays out.

These different aspects have been extensively theorised over a long historical trajectory dating back to the ancient Greeks and in this chapter we offer a situated approach to complex ideas, grounded in the main interpretations from philosophers and theorists from geographical backgrounds.[4] Philosophers have been very aware of changing ways of understanding the physics of time, and the significance of *time frames* adopted for the measurement or classification of time as an external metric. From the paradoxes explored by Zeno, through Newtonian mechanics, to the relative challenges of Einsteinian thought and quantum physics, scientific approaches to time as something outside of human experience continue to be profitably explored. Arguably though, the most relevant philosophical debate has focused on *temporality* as 'lived' or 'experienced' time, contrary to time theorised by physicists (see Hoy, 2012). The ways in which people apprehend and experience the temporal is of central concern for our arguments about digital mapping. May and Thrift (2001) suggest that four different domains characterise this geographical apprehension of the temporal: a focus on natural rhythms and cycles, on systems of social discipline, on devices and technologies, and on conceptualisations. Our authors' concerns with Adam's aspects of time play out in lived human contexts; where digital mapping technologies evoke, but also themselves contribute to, making temporalities in particular social contexts. These studies are frequently underpinned by phenomenological and existential conceptualisations of temporality, or by conceptions of temporality that are strongly influenced by material-semiotic understandings of technologies such as digital mapping. So it makes sense to explore how different foundational thinkers have approached time as a necessary prerequisite for the case evidence.

Concerns for the nature of time are one of the most important aspects of phenomenological thought. Phenomenological approaches to *temporality* regard time as emerging in our individual capacities for making sense of the world. The father of phenomenology, Edward Husserl, focused on exploring what he termed 'internal time consciousness' – suggesting that the temporal was a central and indivisible aspect of being human (Hoy, 2012). This idea of 'being' was developed in Heideggerian thought through hermeneutics. Heidegger makes an important distinction between different modes of temporality, which he characterises as 'ontic' and 'ontological'. Ontic knowledge relates to the properties of things, whereas ontological knowledge 'is the basis on which any such theory (of ontic knowledge) could be constructed, the a priori conditions' (Elden, 2001: 9).

Henri Bergson's work with its focus on *duration* has been strongly influential, across many different disciplines addressing time. Bergsonian ideas regard time as mobile and incomplete, experienced in varying subjective ways and emerging from inner human life. The relation of individual and social memories is also an important theme emergent from Bergson's conceptualisations of temporality, which can be brought to life in digital mapping. These ideas are imbued with vitalism – the notion that some kind of life force beyond science imbues existence (Greenhough, 2010). This emphasis on temporal vitality is reflected in the rise of what has been termed new Bergsonian thought (May and Thrift, 2001), and is crucial for thinkers such as Gilles Deleuze. Deleuzian ideas embody a rhizomatic flowing and folding of temporality. Smith (2013) highlights the ways in which Deleuze freed time, giving it a power to do work, through processes of anticipation, archiving and synthesis. This emphasis inevitably also focuses attention onto *timing* and the momentary event. The non-representational theory that emerged in the middle of the 1990s as a post-phenomenological response to inject life into the dead geographies of representation (Thrift and Dewsbury, 2000), also regards time as contingent, and has focused on practice and performance as immanent and embedded in actions (Lorimer, 2005; Anderson and Harrison, 2010). Anticipation, pre-emption and pre-caution characterise much recent geographical scholarship in this vein (see for example Amoore, 2013).

In Marxist material readings of change, documenting time-space compression (*tempo*) has been read as an outcome of changing modes of production (see Harvey, 1989). Technologies have clearly been deployed to speed up processes and in so doing, facilitate new cycles of capitalist accumulation. In extreme accounts, such as those of Paul Virilio, space itself becomes irrelevant in the acceleration occasioned by technological advance and where tempo changes everything (May and Thrift, 2001).

The *sequence* of time and predictability has also generated significant debate. Derridean deconstructionism distinguishes between the inevitability of *'futur'* – the future that is foreseeable and programmed – as against the impossibility of predicting, which he terms *'l'avenir'* – that which might come. The contrast between an animated inevitable outcome and a stochastic, unknowable other presents significant challenges addressed in several of our subsequent chapters. Futurity, for Derrida, remains open yet structured by history, and we only access this through events that are yet to come about (Hodge, 2007). A concern for genealogy and historicised interrogation of subject positions also informs Foucauldian notions of temporality, in which future power relations are disciplined by present and rational regimes of governance and biopower (Hoy, 2012).

There has been a blurring of the organic and inorganic in many recent approaches to the *modalities* of time, evidenced particularly in work drawing on traditions of science and technology studies (May and Thrift, 2001). A focus on digital mapping inevitably draws on the materiality of maps as 'things': on screens, as code, and embedded in other objects that circulate with varying degrees of mutability, freezing time, but also animating other human action. Actor-network explorations of temporality foreground the relations between technologies and other actants with an ongoing capacity to change life. The role of time in this kind of material assemblage is emergent and contingent; instead of working as a kind of background, time works as situated and is made by everyday practices and relations (Sørensen, 2007). The early 2000s presented a plethora of opportunities to study the digital aspect of such assemblages, given the rise in ubiquity of advanced cartographic displays. Parallel (but often apart from) the discussion in the social sciences and the humanities, cartographers and Geographical Information Science specialists debated over the rising challenges of making time accessible and understandable to increasingly diverse categories of users. The changes in velocity and volume of data, coupled with advanced ways of visualising it, created ample opportunities for experimental approaches to mapping (Kraak, 2003).

*Rhythm* has also been approached from different positions. Parkes and Thrift (1979) were among the first geographers to explore the potential power of considering how timetables and natural rhythms structure space. An academic turn towards researching mobilities further encouraged a focus on the performance of different rhythmic patterns. Influential in this context, also from a Marxist perspective, has been the work of Henri Lefebvre (2004), emphasising what he termed rhythm-analytical approaches to everyday life. More cultural reflections on rhythm and space characterise Tim Edensor's (2012) work.

Imagined relations of space and time, as reflected in Bakhtin's notion of the chronotope, offer another powerful tool for categorising the ways space and time come together in particular mapped configurations. Mike Crang (2001) suggests that the chronotope might usefully be extended beyond its original focus on imagined worlds of the novel, to other forms, such as urban life. The relations of space and time can in this account be charted as a kind of flowing, spatialised temporality, and the chronotope might usefully help to organise understandings of mapping imaginary reconstructions of moments, rhythms, memories, flows and processes. So a richly varied philosophical diversity characterises the ways in which geographical theorists have imagined the temporal, and our authors in this collection also deploy different thinkers in their analysis of the relations of mapping and time.

## Time for a temporal turn

In spite of this long-standing and complex philosophical interrogation of temporality, commentators have argued that at various times from the 1980s, space has become much more significant than time in academic discourse. In part, this is a reaction to grand meta-narratives of Marxist thought (Massey, 2005). Figure 1.1, for example, shows how the term 'spatial turn' has been deployed in the last twenty years in the corpus of books published in the English language, as archived by Google. It highlights a spectacular growth in concern for local and spatial understanding, opposing universal explanations pursuing historical and often structural ways of knowing the world. Earlier concern for space had been demonised as acritical and reificatory, and the widespread acceptance of more partial and local ways of approaching research only really advanced with the widespread adoption of social constructivist and post-structural thinkers such as Said, Foucault, Lefebvre and Derrida. The trend continues to be evidenced in the frequent citation of luminaries of postmodern ways of knowing the world, from Soja, to Dear and Gregory, at varying times in different disciplines across the humanities and social sciences (Warf and Arias, 2009). This continuing

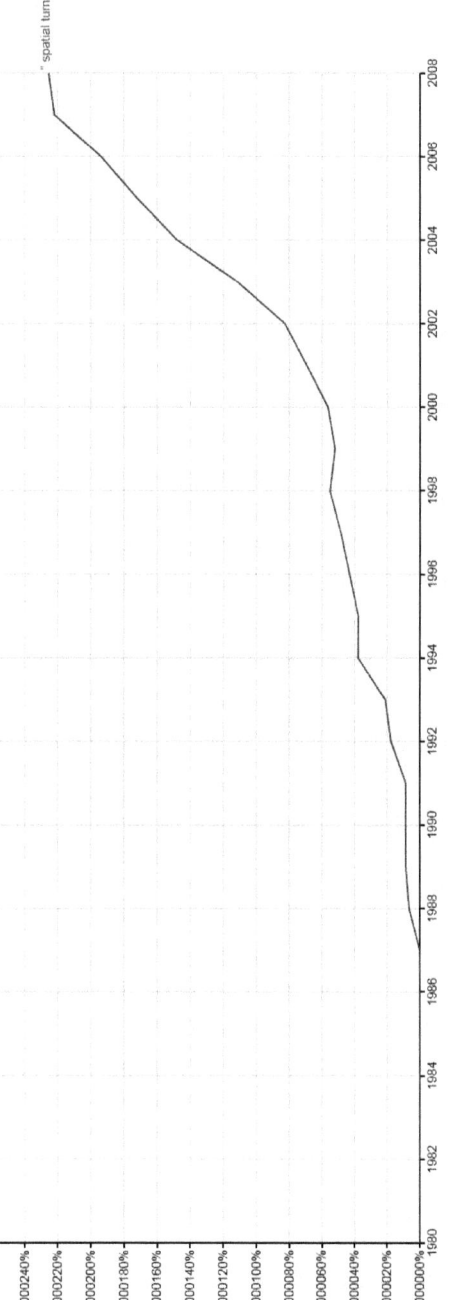

**Figure 1.1** 'Spatial Turn' over time (Google Books Ngram Viewer, http://books.google.com/ngrams).

spatial trajectory has implications for the ways in which temporality has been treated, or rather often elided, in academic work.

In particular, mapping has largely been regarded as a quintessentially *spatial* pursuit during the last thirty years of research and practice. Cognitive cartographic research focused upon *spatial* cognition (Perkins, Kitchin and Dodge, 2011). The history of cartography was safely separated from more mainstream social scientific research. Contemporary research focused on functional improvements in the ability to depict space and design interfaces to communicate knowledge about space. Geographical Information Science focused upon *spatial* analysis. From its inception, the internet was conflated with spatial metaphors such as 'cyberspace' and its topology charted (see for example Dodge and Kitchin, 2001; Graham, 2013). Contemporary digital communication networks facilitate this compression of time and space by 'flattening' the world.[5] Historians only began to deploy mapping technologies to investigate historical processes relatively late. This spatial fixation was critiqued from political and social positions, but interestingly temporality was only rarely explored by critical thinkers focusing on mapping. Further, radical rethinking of space-time and time-space from the position of animating the construct (see Merriman, 2011) underplays the implications of digital mapping for temporality: a point we return to in the conclusion to this chapter. For the moment though, it is safe to argue that few attempts have been made to resolve the paradoxes of this spatio-temporal dualism from the perspective of the *digital* aspect of digital maps. As previously noted, platforms like Google Maps have quickly entrenched themselves in the mundane conduct of users, building on 'social' and 'frictionless' qualities unique to contemporary mediascapes. Yet, when unpacking such objects of inquiry, the temporal is often under-theorised.

New media theorists of the late 1990s and early 2000s, by and large, considered digital modes of expression as extensions or re-mediations of televisual or filmic framings of space (Bolter and Grusin 2000; Manovich, 2001); as a communal space without boundaries for the expression of cultural identities (Baym, 1999; Jenkins, 2006); or as a virtual realm separate from, and/or in flux with, 'common' human biologies (Hayles, 2002). Already a decade ago, communication researcher Mark Nunes, picking up on the unique language of digital engagement used by academics and popular media alike, critically noted that:

> while the rhetoric may have cooled, reference to the Internet in spatial terms – as cyberspace or some other place – still occurs frequently on television and in print. This sense of space is also still mapped by the verbs of displacement – browsing, cruising, going – that have become common parlance for our human–computer interactions, along with our use of locatives and the spatial-geographic language of sites, addresses, and links that describe the material, conceptual, and experiential arrangements of

the World Wide Web and other instances of networked communication. (Nunes, 2006: 2)

If the internet is firmly anchored in spatiality and its language, what hope did the 'geoweb' and closely related genealogy of digital cartography ever have in escaping similar framings?

Della Dora (2012) notes how the point of departure for contemporary digital mapping imaginations are embodied in US Vice-President Al Gore's famed 'Digital Earth' speech from 1998, in which he envisioned a joint mega-project of a singular virtual globe, maintained by multiple organisations and accessible to all. Such a creation would have been used for tasks as diverse as fighting crime, conducting diplomacy and preventing climate change. While lacking the sophistication of Gore's brainchild, modern digital maps are nonetheless engulfed in the same ethos of all encompassing 'truthful' territoriality. The likes of Google Maps, packaged with such services as 'My Places' or 'Street View', promise a convenient convergence of localities onto our multiple screens. At the same time, contemporary computational regimes foreground time and its management through the appropriate interfaces of digital media. From personalised phone assistants – like Apple's Siri or Google Now – to the influx of 'productivity' software that aims to help users maximise efficiency in regard to their own time management, guiding and affecting user temporalities has become the 'go-to' way for connecting people to their surrounding environment. To complete our understanding of digital cartography, we must examine it as part of a digital landscape that enacts temporal practices *along with* spatial ones.

The aim of this book is to highlight these practices, as now is the time for a temporal turn across a multitude of disciplines that address digital cartography. With mapping applications approaching their post-digital disenchantment (Cramer, 2014), and no longer viewed as the domain of highly specialised professionals, we see multiple avenues available for exploring such a turn.

First, one can argue that digital – as opposed to paper – mapping interfaces enact a different type of engagement with temporalities. Unlike the envisioned compressed real-time space of flows presented by the iconic world maps divided into time zones, the modern map is intimate and personal. The map at your fingertips is more likely to register your personal histories of travel, shopping or socialisation than it is to tell stories of shared and structured timelines across divides (Hind and Lammes, 2015). It moves as you move, animating time in ways that were impossible in the pre-digital age. Incorporation of asynchronous technologies with growing coverage removes the need (or desire) for a flattening of spaces; allowing such practices as geolocative games, where users incorporate revisiting the same places over time as part of their travel or leisure routines. Wearable technologies, such as smart watches and health-trackers, furthermore maintain personal time zones by allowing users to track their own movements through time and space; to gauge and evaluate them. At the same time, they allow others to monitor personal time, even outside the clock-punching of the school or workplace, reflecting self-governmentality present in societies of control (Deleuze, 1992). Third, the digital allows new analytical depths of existing maps, for example through new re-imaginings, different visualisation techniques, renderings and the like. Such is the example of Minard's famous map of the Napoleonic Russian campaign, re-examined through digital technologies (Kraak, 2014).

These new forms of temporalities are incorporated into activities that go beyond the map *per se*, such as the precise quantification of subjective experiences of wayfinding, presented by GPS SatNav devices, or the almost-gameful 'swiping' experience of casual romance on Tinder, for the off chance of locating suitable partners in the condensed time-space bubble the app affords. New temporalities of working are called into play by the operation of the interfaces of apps such as Uber, or Deliveroo, and are enacted in the precarity and anxiety of the new 'gig' economy (Friedman, 2014). These new forms of temporalities are also embedded in the spell of 'real-time big data', in the rhetoric of the 'smart city' and in the dreams of control that are called into play (see de Lange, this volume).

Digital mappings are also central to an ongoing and increasing interest in 'map art', charted by several of our authors in this collection. By focusing upon new ways of imagining the temporal, that frequently work in opposition to the notion that mapping is useful or instrumental, these artistic re-mappings evoke diverse possible futures, instead of working to anticipate or close down possibilities.

Additionally, temporal considerations are required in order to *map the digital* rather than only *digitally map*. In an ever-changing landscape of companies, practices, technologies and users, the misleadingly simple term of 'mapping' something requires a way to document both *what* a thing is, and *where* it resides, but also *when* and for *how long*. The Livehoods project (Cranshaw *et al.*, 2012; Mobile Commerce Lab, 2015), where temporary maps of modular neighbourhoods in large cities are created based on characteristics of their residents' media use, is one example for such an approach.

New forms of 'liveness' are being called into effect by the technological assemblage around social media (Van Es, 2016). But, despite an intensifying narrative around 'real-time' digital capabilities, little attempt has been made to unpack such claims to technological immediacy. Also, despite routinely hyperbolic promises of 'live' data streams, 'background' updates and 'reflexive' systems, most digital users are faced with intermittent or incomplete data streams, intrusive or unwarranted updates and largely unresponsive or generally clunky systems. Each of these promises to generate a rather contrasting vision of technological use. Take, for example, Google's launch of its Maps incarnation in 2014. Unlike two years previously, when Apple's equivalent resulted in a catalogue of geospatial errors – from mislabelled locations to melting photo imagery – Google suffered a different problem. Users began to complain it was slower at executing vital tasks – like loading map data and calculating route options – than the previous version. In response, Google offered a number of solutions: either users try its 'Lite mode' designed for less powerful machines, or upgrade their browser and the various plugins required to optimise in-built functions such as Street View. Although technical solutions of a kind, neither resolved a more resolute conceptual issue: that real-time technological functionality is still just a wild – if not wholly impossible – dream. The instant experiences that are promised to huge fanfares at seasonal product launches, promoted in global advertising material, and regurgitated wholesale by (usually) unwitting brand ambassadors, rarely constitute the reality.

While digital mapping platforms, applications and browsers are valued for their spatial processing capabilities, it is their temporal features that are felt most acutely – especially when they fail. It is failure that so often constitutes the contemporary digital mapping experience. Yet, like the real-time narrative, it has rarely been attended to. The emergence of new temporalities – borne out of the design of various mapping platforms – has led to a suite of novel 'failure spaces' and 'failure types'. Moreover, a plethora of work-arounds, patches, temporary fixes and wholesale reboots have been developed to combat such issues.

## Emergent themes

The chapters in this book show that digital mapping needs to be viewed from an interdisciplinary angle, so as to best grasp what it means in terms of temporality. We have already established that temporality has been explored from many different academic perspectives. Furthermore, multiple positions, methodologies and assumptions are needed to do justice to the nature of digital mappings; being at once creative practices, media, cartographies and technologies. The rich interdisciplinary background of the authors allows these different positions to be developed. We need new ways of knowing and discovering, so as to capture a more nuanced and sensitive view of digital mapping as a phenomenological, cultural and political practice that goes beyond the technical, descriptive and cartographic analysis that predominates in digital mapping research. Through these different ways of knowing, more can be revealed about temporality and its relations to the unfolding worlds of digital mapping. This volume then acts as a kind of temporal wayfinder, showing how temporality matters in contemporary discussions and practices relating to digital mapping and geolocative media, but also flagging some of the ways this might become increasingly important in the future.

We have chosen to divide this book into three parts that are not so much ordered by type of case, or disciplinary anchoring of authors, but are rather grouped so that each part highlights an important quality of temporality in relation to mapping.[6] All contributions move beyond the descriptive to pay particular attention to what we call the 'critical dynamics' of time. In each case contributions focus on, or cross-cut between, digital maps, digital mapping or digital locative media.

## *Ephemerality/mobility*

Contributions in the first part, 'Ephemerality/mobility', consider the fleeting, dissipating and transient aspects of digital mapping and temporality. Whether the process involves making maps or traversing them, these temporal notions of digital mapping are experienced through interweaving narratives of the past and present that cannot always be easily captured. Movement, fluidity and dynamism characterise these different contributions, and they weave together inspiration from Deleuzian, non-representational or Foucauldian thought, with grounded and often performative attention to situated everyday experiences of temporality. The unstable elements of Adam's (2008) typology are emphasised here: the tempo, rhythms, modalities and social practices of temporalities, through which mapping is performed. Each chapter in this part deals with the ephemeral and mobile aspects of the relationship between temporality and digital mapping from a different angle, throughout all the phases of the mapping process, from inception to deployment and user manipulations of the map. By beginning this volume with this particular focus, we encounter digital mapping and temporality at its most personal and situated – whether it be the impending arrival of a GPS unit, the relationship between whispering scents and memory, the personal cartographies of communication, or the disjointed and affective reaction to unexpected glitches.

In 'Nodes, ways and relations', Joe Gerlach spins together a number of intersecting temporalities, narrating the otherwise irretrievable process of producing an OSM map with a group of other volunteer mappers. As a self-described 'intervention', these narratives unfold, deconstruct and liquidise traditional spatial metaphors through a series of animated interjections – Gerlach offers an auto-ethnographic exploration of mapping focusing on the banality of logistics and equipment, while reflecting on the ideological and discursive nature of digital cartography. Together, these threads playfully 'undo' traditional conceptualisations of how mapping and cartographic processes are undertaken and narrated. In so doing, the chapter highlights the contingent, banal and embodied aspects of the ways in which the mapping of time plays out.

The next chapter, a trialogue between graphic designer and 'smell mapper' Kate McLean, Chris Perkins and Sybille Lammes, explores ways in which transient olfactory phenomena may be mapped and how this changes our conception of maps. McLean's mapping practice focuses on the fleeting temporalities of smell mapping. She describes the menagerie of sensory urban traces as quixotic – moving and sweeping over the city, sometimes lingering and at other times, dissipating. Urban smellscapes are deeply temporal, with rhythmic and durative qualities that are often subjective and ambiguous. Accordingly, McLean's digital mapping work aims to capture the movement, transient memories and ephemerality of her subject. Much like Gerlach, McLean's work is an intervention into traditional cartography, which has historically been bounded by static, spatial markers rather than the evanescent sensorium of urbanity and mobility, but it also evokes the mapping processes of the artist, whose practice has itself changed through time.

In the final chapter of this part, Pablo Abend explores how asynchronous temporalities may be animated and enfolded through digital mapping. Here, the passing times and changing seasons are unintentionally melded together in Google Street View – a winter's day from one perspective, a midsummer's night from another – in an algorithmic accident that echoes the poly-temporality of Gerlach's philosophical intervention. Rather than spatialising how Google Street View may appear at different times, Abend begins by drawing on media archaeology (see Parikka, 2012) in order to historicise this fettered relationship between movement (down a street in Street View) and photographic and filmic media (in the form of geographic imagery). Yet Abend argues that the heterogeneity of temporality augers a heterogeneity of 'frames', or singular spatial representations in Google Street View. Through the ephemerality of the seasons the temporal process at work is revealed. Seasons are unique and fleeting, and yet, like the smellscape in McLean's work, the echoes left after their brief appearances can be uncovered.

Thus, 'Ephemerality/mobility' ends on a slightly different note from which it started. It begins by discussing the transience of digital mapping processes and ends with the way these processes are embedded in maps themselves. Furthermore, the embodied mobility in both Gerlach's and McLean's chapters on mapping processes becomes a different kind of mobility – a passage defined by the ephemeral temporalities of digital mapping.

## Stitching memories

The chapters in this part of the book look at various visual practices that engage digital technologies to stitch moments together in a non-linear fashion, in particular, as a way of remembering and capturing time. In so doing, they draw attention to the potential of digital mapping to reveal temporal juxtapositions, and (a)synchronicities. These go against the ephemerality that is so prominent in digital mapping, which is addressed in the first part of the book, not by denying its existence or experience, but by turning attention to both 'creative' *and* 'everyday' practices that counter digital mappings' fleeting instantaneity. So they draw out the implications in particular of Adam's (2008) ideas of time frame, sequence and duration, while also attending to the practices through which an apparent fixity might be constructed.

The first chapter by Rachel Wells, on the work of Wolfgang Weileder, shows how his photographic and other site-specific visual work seeks to undo the transience highlighted in the previous part. It establishes this by visually capturing time and recombining it in images. In his *atlas-project*, he horizontally stacks strips of photographic material shot at exactly the same place, over a stretch of time, thus creating time maps. Wells analyses this slicing and joining of temporal images from a Benjaminian perspective, arguing that Weileder's re-ordering of the ephemeral frees photography of its presumed stillness and reproducibility. Weileder's oeuvre juxtaposes time and expresses a 'dissatisfaction with the time-space coordinates of photography' and a 'withering away of aura' (Wells, this volume).

In his chapter about locative art practices, Gavin MacDonald also talks about recombining mapping temporalities; challenging, or at least problematising, like Wells, the conception of images as fleeting moments. MacDonald argues that this tendency to conceive of digital maps as a-temporal is rooted in two different a-temporalities at work in geomedia: a cartographic suturing of time and space – also redolent of Weileder's *atlas-project* – and a-temporalities generated by the acceleration and immediacy of digital media. To respond to this atemporal tendency he, unlike Wells, brings in the artist at work, showing how Street View captured two locative artists *during* their practice; at once highlighting different moments and re-ordering time. He also draws attention to the temporality of the locative art movement itself, stressing that changing technologies and their everyday deployment have made this discourse obsolete, or as we state earlier 'the digital is the normal' now (Wells, this volume). This is an important observation as it also stresses that possible relations between mapping interfaces and their temporal affordances are themselves changeable and time-bound. As we write, digital mapping has already developed minor histories (Gerlach, 2015), as MacDonald shows when he looks back at the heydays of the locative art movement.

While both previous authors show how cartographic art can be used to map temporalities and meld temporal experiences – in so doing offering opportunities to remember and reconfigure cartographic time – the final chapter in the part is anchored in more 'everyday' practices. In Matthew Hanchard's contribution, 'practice theory' is put to work in order to shine a light on lived, embodied mapping practices. Through an array of cases – interviews with hotel owners and customers, local authority staff and hiking society members – Hanchard weaves a series of rich cartographic stories, drawing attention to the effects that digital mapping has had on both work and leisure practices. He thus draws attention to their temporal impact – on attention, on security and trust, on efficiency and on memory – as well as the methodological possibilities of deploying practice theory to investigate these.

## *(In)formalising*

In the final part of the book, each chapter focuses on how digital mapping *(in)formalises* time. This tension – between the formal and informal – works to delineate how far digital maps have been able to capture, structure and regulate the temporal. Some technological agendas have worked to secure against future instabilities by exercising an anticipatory logic, such as weather maps, disaster maps or military maps (Monmonier, 2010). Others have actively worked to open and extend such a possibility, proliferating and intensifying rather than controlling or eradicating the temporal. In particular situations, digital maps have been designed to set temporal coordinates and codify gestures and actions, for example the Livehoods project mentioned above, which re-maps neighbourhoods based on social media usage of residents. Yet there continues to be a retaliatory strand of ideas, through which digital mapping re-works coordinates, rendering them unreliable, 'fuzzy' or unknown altogether. The following chapters challenge the inexorable march of dominant tempo-spatial linguistics, narratives and discourses by consistently returning to digital mapping's (in)formalising tendencies. Hence they each emphasise different aspects of Adam's (2008) trope of temporality, highlighting the tensions between the institutional and the personal and between the regulated and the open.

In Thomas Sutherland's chapter – 'Mapping the space of flows' – the notion of the 'flow' and its cartographic materialisation as a 'flow map' is put under the spotlight. Taken as a 'natural and unproblematic way of describing the temporalities and mobilities of digital, networked capitalism' (Sutherland, this volume), the notion of the flow risks reifying capital's apparently inevitable fluid advance. The flow map, in turn, becomes the carrier and modulator for this double threat, presenting the world as an unceasing whirlwind of goods, services, capital and labour. Yet, as Sutherland suggests, the flow map has a much longer history, with its roots traceable to itinerant documents such as the *Tabula Peutingeriana*.[7] Nineteenth-century trade and battle maps crafted to show the extent of colonial power, however, arguably represent the true 'birth' of the genre. Yet, these flow maps have not only persisted, but have proliferated. There is, it seems, a rather unhealthy obsession with rendering the world as full of flows. The ironic repercussion of maintaining this obsession, Sutherland argues, is a complete ignorance of the turbulent nature of global life, replete with stoppages, blockages and fractures. Scripting an *a priori* form to such 'flow-thinking' – and by extension, the flow map – risks generating a world in which flow is seen as both normal and ontologically apparent.

Yet as Cate Turk makes clear in her chapter – 'Maps as foams and the rheology of digital spatial media' – the world is far from ontologically apparent. Despite a shift in the cartographic exhibition of phenomena – from static 2D to dynamic 3D maps – this new-found dynamism is missing from many contemporary analyses. More attention needs to be paid to the different types of dynamic features found within digital maps: from slippy interfaces and automated location preference features, to 'live' data feeds and evolving software updates. But as well as conceptual concerns, the chapter also draws attention to the methodological implications of dealing with dynamism in digital maps, by asking whether 'freezing' phenomena is necessary in order to analyse them. In an attempt to avoid such an icy approach, Turk novelly applies Peter Sloterdijk's work on bubbles and foams (see for example, Sloterdijk, 1998; 1999; 2004) to the world of humanitarian crisis mapping, in order to better comprehend their contingent and temporally variable ontological qualities.

Tuur Driesser is equally concerned about the ontological implications of new digital mapping platforms. His contribution – 'Maps as objects' – addresses the ways in which a 'pathogen weather map' of New York City, called PathoMap, articulates the relationship between present and future visions of the city, by mapping its 'microbial population'. Driesser suggests that city officials provisionally become able to prepare for, mitigate against and recover from an array of national and anthropogenic disasters. Concomitant with broader narratives around anticipation, contingency and foreclosure of future action, the PathoMap marks the shift from a politics of risk, to a 'politics of the possible' (Amoore, 2013), in which future threats (such as pandemics) are secured against.

Michiel de Lange's chapter – 'From real-time city to asynchronicity' – is an equally ambitious and provocative explication of a new cartographic vision of the city. In expanding on Bleecker and Nova's (2009) call for an 'asynchronous' urban narrative, de Lange challenges the notion of 'real-time' that writes serendipity and the 'everyday messiness' out of city life. Urban dashboards – used by state departments and local authorities – further this 'realist epistemology' (Kitchin, Lauriault and McArdle, 2015: 13; Mattern, 2015) in which access to urban phenomena is unproblematically assumed to be both possible and reliable. The 'asynchroni-city', as proposed by Bleecker and Nova and taken up by de Lange, is a speculative – but not altogether hopeless – attempt to re-direct the dominant, neoliberal narrative away from an 'efficiency-driven real-time model' that favours a calculable and known future, to ask whether we should argue for a 'slow mapping' of the city. In critiquing an array of 'smart' urban dashboards, de Lange emphasises how asynchronous design 'may contribute to a "social" alternative' of adaptive, mutable, cartographic platforms.

## Foreclosing the future

Temporality offers an important framework for understanding digital mapping in all its diverse forms. Deploying a temporal lens delivers fresh perspectives for understanding and analysing mapping practices, which is indispensable for understanding the form and affordances brought together in digital mapping. It also draws attention to different knowledge claims about time, temporality, time-space and space-time.

At the start of this chapter we introduced Merriman's (2011) concerns with non-digital animation and his rejection of static notions of spatio-temporality and tempo-spatiality. By focusing on ephemeralities, stitching and (in)formalities, the chapters in this book highlight the transformational potential of a temporal turn that moves beyond Merriman's concerns, taking on board different situated examples of digital mapping practice. Building on his conceptual idea of movement, our authors show how aspects of mapping practice might be understood as material, but also more-than-representational, socio-cultural phenomena. By focusing on a rich diversity of cases, from OpenStreetMap, to PathoMap and urban dashboards, the links between Cartesian ontologies and multiple temporalities are foregrounded. We are still governed by temporalities that are grounded in ideology, and discourse (pace; Merriman, 2011), but our chapters also enrol particular situated and temporal contexts.

So, the chapters in this book all take up Merriman's call for a rethinking of temporalities and apply this to the worlds of digital mapping. Qualities of digital mapping begin to emerge in these stories, that would otherwise have remained under-theorised.

## Notes

1 We regard spatio-temporalities as starting from space: tempo-spatialities in contrast start from a consideration of the temporal.
2 The fictional setting of the *Back to the Future* franchise.
3 See www.seeing-stars.com/Locations/BTTF-Map.shtml (accessed 21 November 2017).
4 Geographers have a particular fascination with mapping (see Dodge and Perkins, 2008), but also with temporality (see May and Thrift, 2001).
5 A term popularised by the *New York Times* columnist Thomas Friedman.
6 The editorial team arrived at this tripartite division after an extended coding session, discussing similarities and differences between chapters.
7 The *Tabula Peutingeriana* illustrates the road network of the fourth- or fifth-century Roman Empire.

## References

Adam, B. (2008) Of timespaces, futurescapes and timeprints. [Online] Available at: www.cardiff.ac.uk/socsi/futures/conf_ba_lueneberg170608.pdf (accessed 1 August 2016).
Amoore, L. (2013) *The Politics of Possibility: Risk and Security Beyond Probability*. Durham, North Carolina: Duke University Press.
Andersen, C. U., Cox, G. and Papadopolous, G. (eds) (2014) *APRJA Post-Digital Research Special Issue*, 3(1). [Online] Available at: www.aprja.net/?page_id=1291 (accessed: 3 August 2016).
Anderson, B. and Harrison, P. (eds) (2010) *Taking Place: Non-Representational Theories and Geography*. London: Routledge.
Baym, N. K. (1999) *Tune In, Log On: Soaps, Fandom, and Online Community*. Thousand Oaks, California: Sage Publications.
Berry, D. and Dieter, M. (2015) *Postdigital Aesthetics: Art, Computation and Design*. London: Palgrave Macmillan.
Bleecker, J. and Nova, N. (2009) 'A synchronicity: Design fictions for asynchronous urban computing', *Situated Technologies Pamphlet* 5. New York: The Architectural League of New York.
Bolter, J. D. and Grusin R. (2000) *Remediation: Understanding New Media*. Cambridge, Massachusetts: The Massachusetts Institute of Technology Press.
Bratich, J. (2006) Public secrecy and immanent security: A strategic analysis. *Cultural Studies*, 20(4–5): pp. 493–511.
Cramer, F. (2014) What is 'post-digital'? *APRJA* 3(1). [Online] Available at: www.aprja.net/?p=1318 (accessed: 3 August 2016).
Crang, M. (2001) 'Temporalised space and motion'. In: May, J. and Thrift, N. (eds) *Timespace: Geographies of Temporality*. London: Routledge, pp. 187–207.

Cranshaw, J., Schwartz, R., Hong, J. I. and Sadeh, N. (2012) *The Livehoods Project: Utilising Social Media to Understand the Dynamics of a City*. Proceedings of the Twenty-Sixth International Association for the Advancement of Artificial Intelligence, Toronto, Canada, 22–26 July.

Deleuze, G. (1992) Postscript on the societies of control. *October*, 59: pp. 3–7.

Della Dora, V. (2012) A world of 'slippy maps': Google Earth, global visions, and topographies of memory. *Transatlantica American Studies Journal*, 2. [Online] Available at: http://transatlantica.revues.org/6156 (accessed 3 August 2016).

Dodge, M. and Kitchin, R. (2001) *Atlas of Cyberspace*. London: Addison-Wesley.

Dodge, M. and Perkins, C. (2008) Reclaiming the map: British geography and ambivalent cartographic practice. *Environment and Planning A*, 40(6): pp. 1271–1276.

Edensor, T. (ed.) (2012) *Geographies of Rhythm: Nature, Place, Mobilities and Bodies*. London: Ashgate Publishing.

Elden, S. (2001) *Mapping the Present: Heidegger, Foucault, and the Project of a Spatial History*. New York: Continuum.

Friedman, G. (2014) Workers without employers: Shadow corporations and the rise of the gig economy. *Review of Keynesian Economics* (2): pp. 171–188.

Gerlach, J. (2015) Editing worlds: Participatory mapping and a minor geopolitics. *Transactions of the Institute of British Geographers*, 40(2): pp. 273–286.

Graham, M. (2013) Geography/internet: Ethereal alternate dimensions of cyberspace or grounded augmented realities? *The Geographical Journal*, 179(2): pp. 177–82.

Greenfield, A. (2006) *Everyware: The Dawning Age of Ubiquitous Computing*. San Francisco, California: New Riders.

Greenhough, B. (2010) 'Vitalist geographies: Life and the more-than-human'. In: Anderson, B. and Harrison, P. (eds) *Taking Place: Non-Representational Theories and Geography*. London: Routledge, pp. 37–54.

Harvey, D. (1989) *The Condition of Postmodernity: An Enquiry into the Origins of Cultural Change*. Oxford: Blackwell.

Hayles, K. N. (2002) Flesh and metal: Reconfiguring the mindbody in virtual environments. *Configurations*, 10(2): pp. 297–320.

Hind, S. and Lammes, S. (2015) Digital mapping as double-tap: Cartographic modes, calculations and failures. *Global Discourse*, 6(1–2): pp. 79–97.

Hodge, J. (2007) *Derrida on Time*. London: Routledge.

Hoy, D. C. (2012) *The Time of Our Lives: A Critical History of Temporality*. Boston, Massachusetts: The Massachusetts Institute of Technology Press.

Jenkins, H. (2006) *Convergence Culture: Where Old and New Media Collide*. New York: New York University Press.

Joliveau, T. (2009) Connecting real and imaginary places through geospatial technologies: Examples from set-jetting and art-oriented tourism. *The Cartographic Journal*, 46(1): pp. 36–45.

Kitchin, R., Lauriault, T. P. and McArdle, G. (2015) Knowing and governing cities through urban indicators, city benchmarking and real-time dashboards. *Regional Studies, Regional Science*, 2(1): pp. 6–28.

Kraak, M. J. (2003) *The Space-Time Cube Revisited from a Geovisualisation Perspective*. Proceedings of the 21st International Cartographic Conference, Durban, South Africa, 10–16 August. [Online] Available at: http://icaci.org/files/documents/ICC_proceedings/ICC2003/Papers/255.pdf (accessed 1 August, 2016).

Kraak, M. J. (2014) *Mapping Time: Illustrated by Minard's Map of Napoleon's Russian Campaign of 1812*. Redlands, California: Esri Press.

Lefebvre, H. (2004) *Rhythmanalysis: Space Time and Everyday Life*. Translated by S. Elden and G. Moore. London: Continuum.

Lorimer, H. (2005) Cultural geography: The busyness of being 'more-than-representational'. *Progress in Human Geography*, 29(1): pp. 83–94.

Manovich, L. (2001) *The Language of New Media*. Cambridge, Massachusetts: The Massachusetts Institute of Technology Press.

Mason, P. (2015) *Postcapitalism: A Guide to our Future*. London: Macmillan.

Massey, D. (2005) *For Space*. London: Sage Publications.

Mattern, S. (2015) Mission control: A history of the urban dashboard. *Places Journal*, March. [Online] Available at: https://placesjournal.org/article/mission-control-a-history-of-the-urban-dashboard/ (accessed 3 August 2016).

May, J. and Thrift, N. (eds) (2001) *Timespace: Geographies of Modernity*. London: Routledge.

Merriman, P. (2011) Human geography without time-space. *Transactions of the Institute of British Geographers*, 37(1): pp. 13–27.

Mobile Commerce Laboratory, Carnegie Mellon University (2015) *Livehoods*. [Online] Available at: www.livehoods.org/ (accessed 29 May 2016).

Monmonier, M. (2010) *No Dig, No Fly, No Go: How Maps Restrict and Control*. Chicago, Illinois: University of Chicago Press.

Ní Fhlainn, S. (2010) *The Worlds of Back to the Future: Critical Essays on the Films*. Jefferson, North Carolina: McFarland and Co Inc.

Nunes, M. (2006) *Cyberspaces of Everyday Life*. Minneapolis, Minnesota: University of Minnesota Press.

Parikka, J. (2012) *What Is Media Archaeology*. Cambridge: Polity Press.

Parkes, D. and Thrift, N. (1979) Time spacemakers and entrainment. *Transactions of the Institute of British Geographers*, 4(3): pp. 353–372.

Perkins, C., Kitchin, R. and Dodge, M. (2011) 'Introductory essay: Cognition and culture'. In: Dodge, M., Kitchin, R. and Perkins, C. (eds) *The Map Reader: Theories of Mapping Practice and Cartographic Representation*. London: Wiley, pp. 1–26.

Sloterdijk, P. (1998) *Sphären I – Blasen*. Frankfurt: Suhrkamp.

Sloterdijk, P. (1999) *Sphären II – Globen*. Frankfurt: Suhrkamp.

Sloterdijk, P. (2004) *Sphären III – Schäume*. Frankfurt: Suhrkamp.

Smith, D. W. (2013) Temporality and truth. *Deleuze Studies*, 7(3): pp. 377–389.

Sørensen, E. (2007) The time of materiality. *Forum Qualitative Sozialforschung/Forum: Qualitative Social Research*, 8(1). [Online] Available at: www.qualitative-research.net/index.php/fqs/article/view/207/457 (accessed 01 August 2016).

Srnicek, N. and Williams, A. (2015) *Inventing the Future: Postcapitalism and a World Without Work*. London: Verso Books.

Thrift, N. and Dewsbury, J. D. (2000) Dead geographies – and how to make them live. *Environment and Planning D: Society and Space*, 18(4): pp. 411–432.
Van Es, K. (2016) *The Future of Live*. Cambridge: Polity Press.
Warf, B. and Arias, S. (eds) (2009) *The Spatial Turn: Interdisciplinary Perspectives*. London: Routledge.

# Part I

Ephemerality/mobility

# 2

# Nodes, ways and relations

*Joe Gerlach*

### Here, now

Maps, mappings, cartographies; (dis)orientations for the everyday, obdurate disciplinary motifs of and for geography, maligned and admired in variable measure. Cartography; a science and set of practices once pertaining to sovereign power alone, yet now increasingly diffuse in its geographic reach and performance. Nonetheless, whether rendered through hegemonic, quotidian or hybrid assemblages, mapping remains resolutely (geo)political at a range of disparate registers; statist to somatic. Elsewhere, I have used three cartographic attributes, namely those of 'lines', 'contours' and 'legends' as a means of establishing a conceptual vocabulary that underscores the awkward, non-representational aspects of a recent surge in participatory, or 'vernacular' mapping practices themselves propelled by the mass-mediatisation of cartography via mobile, digital and online platforms (see Gerlach, 2014). Here I appropriate a further three attributes; drawn from both the cartographic-lexicon-at-large and from the toolkit of OpenStreetMap, the increasingly renowned wiki-based, crowd- and open-sourced mapping organisation established in 2004. These three attributes, *nodes*, *ways* and *relations*, act simultaneously as both the very constitutive material of OpenStreetMap's cartographic output and likewise as points of meditation in this chapter. Together they animate the empirical upshot of the burgeoning conceptual matter on mapping that itself is steadily animating the processual and ontologically/epistemologically insecure natures of contemporary cartography (Kitchin and Dodge, 2007; Crampton, 2010; Elwood and Leszczynski, 2013; Burns, 2014; Caquard,

2014; Leszczynski, 2014), not least highlighting its radical mutability in time and space (Perkins, 2014).

Centred on a chronologically ambiguous narration of a day's ambulant cartographic-fieldwork with and for OpenStreetMap in the English town of Witham, Essex, the chapter will illustrate how the production of nodes, ways and relations is as much a conjuring of unsettled temporal and spatial sensibilities as it is a charting of 'known-knowns'. In other words, the act of surveying an already well-mapped settlement is far from a redundant act of mimesis, but a geopolitical act of repetition, renegotiation and re-territorialisation of time-spaces; the cultivation of what Olafur Eliasson describes as a 'micro-sensibility [of space]' (quoted in Jellis, 2015: 372). Moreover, by thinking about time not as chronological, but as 'durational', after the French philosopher Henri Bergson, it becomes possible to witness the peculiar rhythms of mapping as generative of vernacular and micropolitical space-times. But first, a very brief reprise of contemporary cartography and a word, or so, on vernacular mapping.

## Cartography, unbound

> I am in an unfamiliar place … momentarily what the map tells me bears little sense of where I think I am. (Newling, 2005: 48)

Maps and mappings maintain a curious hold on everyday geographies. On the one hand they enable a *slowing down* of time-space for navigation, thereby (sometimes) quickening journeys to pre-figured destinations. Moreover, both cartographic reason and the cartographic settlement have helped prise apart a modern constitution of ubiquitous dualisms (Latour, 1993) comprising: rifts between representation and reality, between subject and object, nature and culture, between body and map and indeed between time and space; the geometric logic propagating taxation, territory, terror, war and frontier (Lacoste, 1973). Yet on the other hand, as Newling's bewilderment attests to, maps and mappings continue to disorientate and perplex. The chronologically ritualistic choreographies between bodies and cartographies, between satellite navigation devices and world-weary drivers, for example, manifestly complicates any straightforward division between maps and their users, or between representation and the onflow of reality. At the same time, as enunciated by Nigel Thrift (2012), there is a pandemic underway in mapping technologies and practices, which themselves have recalibrated the natures and geographic character of cartographic performances (Chambers, 2006). This has constituted a move away from a sovereign or statist accord of mapping and a shift instead towards a more, broadly speaking, *participatory* model of cartography, facilitated to a large extent by the recent proliferation in Web 2.0; an 'epoch' of the internet dominated by online interfaces, interactive platforms and protocols (Zook *et al.*, 2004; Zook and Graham, 2007). In calling for a 'cartography unbound', a mapping untethered from the stultifying impulses of representational and temporal certitude, a certain degree of conceptual and empirical disorientation and disruption will be inevitable, such are the vernacular energies required to do the work of unhinging cartography from its determinedly Euclidean grid. The upshot, however, is a re-orientation of sorts; an understanding of cartography that valorises its affective, virtual and performative potential, a speculative mapping that questions not what maps 'reveal', but instead anticipates what kind of worlds and geographies maps could bring into being; this is at the core of the idea of 'vernacular mapping'.

Vernacular mapping acts, conceptually, to accentuate the role of bodies in the folds of experience and to diagram the importance of bodies, human and non-human, affective and virtual, in the distribution and performance of everyday cartographies. It is to accept at all times and spaces that the world is never presented in advance, but that maps and their practitioners are always provisional, subject and vulnerable to ongoing modification and change. The topology of mapping is therefore not the preserve of mental cognitive processes, but a distributed set of bodily, way-finding (dis)orienting performances; a cartography for moving spaces. To speak of vernacular mapping is to shake off the 'geometric habits that reiterate the worlds as a single grid-like surface open to the inscription of theoretical claims or uni-versal designs' (Whatmore, 2002: 6), and is instead to diagram a vibrant cartography that is, 'necessarily topological, emphasising the multiplicity of space-times generated in/by the movements and rhythms of heterogeneous association' (Whatmore, 2002: 6).

## Stop, momentarily

'… everything has its geography, its cartography, its diagram' (Deleuze, 1995: 33).

## Fieldwork traces

What follows is a fieldwork tracing of a day's mapping with OpenStreetMap in the town of Witham, Essex, UK. As a then newcomer to the OSM venture, part of the motivation was to learn and acquire the skills and literacies required to 'make it' as a competent OpenStreetMapper, while at the same time reflecting upon the somatic, affective and micropolitical energies of mapping. Methodologically, this is something of a 'fieldwork *of* fieldwork' insofar as a central activity of OpenStreetMap is one of 'surveying the field'; a mode of cartographic interpolation. As such, this modest tracing of a day's cartography joins a growing number of studies that approach examples of vernacular mapping in an ethnographic manner (see Perkins and Dodge, 2008); a realisation of human geographers' tentative re-acquaintance with all things cartographic in the face of calls that suggest cartography is facing an existential crisis.[1]

OpenStreetMap is interesting to geographers in three respects. First, OSM is part of a broader culture of crowd-sourcing whereby the production and marshalling of ideas, knowledges, spaces, and indeed politics, is not carried out by an individual entity or defined by a singular subjectivity, but is instead part of an ecology of practices (Stengers, 2005) in which what matters and what is generated is at risk (Latour, 2008), liable to alteration at any moment by any actor, just as cartographically, OSM is itself at risk. Second, despite the relative novelty of OSM's digital platform, there remain strong resonances with the manner in which cartography has been traditionally conceived and practised; specifically, that the role of mapping is to survey, to abstract, to diagram, to render and to contest what is included and omitted from the map, to argue 'what counts'; a performed typology of cartographic orthodoxy that endures its digitisation. Third, there is something to be said about experience and abstraction. OSM deals in abstractions, but it also deals in experiences insofar as these abstractions are generated in the mapping; that is to say that the digital and analogue experiences tied up in OSM cartography amount to vibrant abstractions, further disrupting the Euclidean representational settlement of abstraction held in distinction from reality, divorced from experience; the myth of maps as mirrors of the world (Rorty, 2009).

In this context, by harnessing the experimental tenor of some cultural geographic writings (see, for example Lorimer and Wylie, 2010; McCormack, 2014), an attempt is made to narrate a story that hangs, counter-intuitively, on the disorientation of mapping, but which finds concomitant re-orientations through nodes, ways and relations. It is a story of cartographic bricolage – of making do (de Certeau, 1984); of mapping subjectivities and time-travelling. There is some degree of cartographic orientation on-hand for guidance; these orientations come in the form of GPS traces and geodesic coordinates. The GPS traces are spatio-temporal riffs, the lived, living and to-be-lived traces of walks, bus and train journeys taken in or near the accompanying geodesic coordinates; namely the marker of where the mapping took place. Known colloquially as 'electronic breadcrumbs', they are the spectral remains of cartographic toil, pseudo-memories and digital inscriptions of an event and its lines, but more importantly, the knitting together of abstraction and experience; a nod to the interstitial space-times of cartography and the crumbling ontological separation of the world and its representation. Some of the traces also signal failure: GPS calibration malfunctions, random lines generated between errant extra-terrestrial satellites and handheld receivers, relations independent of human design, lines, patterns and space-times more broadly emblematic of the setbacks and errors encountered in the performance of vernacular mapping. Similarly, the geodesic markers (longitudinal and latitudinal coordinates) act as spatial refrains, not to anchor the account to any particular locality, but to perform literally as points of departure for the stories, and also to provide resting points for the reader in the same way a map might offer a momentary pause and downward glance from a hiker on the move. Both the GPS traces and geodesic markers are resolutely of a Euclidean vocabulary, but that does not mean necessarily that they must be understood in geometric and cognitive terms alone. These stories, instead, offer a way to think vernacular cartographies through experiential registers.

Witham is a town in Essex, England (Figure 2.1); a reassuringly concrete fact writ-large in the geographic compendium par excellence, the atlas. Witham's cartographic existence is less assured; the rest of it is yet to be mapped, by OSM that is. The GPS handheld device ordered online several days ago has not arrived, and concern is building that I will be turning up to the mapping party on the designated Saturday like a kid arriving at the school gates without uniform. It is bad enough that this will be my first OSM mapping party, previous parties in Maidstone and Haslemere fell by the wayside due to volcanic ash,[2] but apparently the other mappers did a good job. Not total coverage of the Kent and Surrey towns, but nearly there (as reported by TimSC[3] who coordinated the Maidstone mapping party and participated in the Haslemere event). Total coverage? SteveC, founder of OSM, suggested this

**LONG. 0° 38' 11" E; LAT. 51° 47' 60" N**

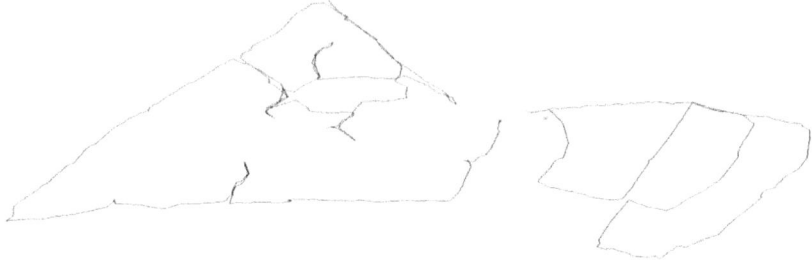

**Figure 2.1** GPS route tracing around Witham, Essex (author's image).

was an ambition OpenStreetMappers needed to divest themselves of, the virtue of a wiki is to always be in the making, never to be complete.

## What is a mapping party?

> A big aspect of getting OSM off the ground was the mapping parties: getting drunk and arguing with people. (SteveC)

A mapping party is an event at which OSM newcomers and veteran enthusiasts gather to map an area of the world that is yet to be mapped, or to edit maps already drawn. They are informal affairs held normally over weekends, organised by a clutch of dedicated OpenStreetMappers, invariably book-ended by milky tea in the morning and warm beer in the evening. In the interim, the gathered mappers walk/drive/cycle to blank areas on the OSM map with GPS devices and/or pens/pencils/paper in hand. They set about walking the streets, recording their tracks and marking waypoints and points of interest along the routes taken; all of which are inscribed through either digital or analogue means, or by a hybrid craft of both. Once the points, tracks and lines have been recorded, the data is then uploaded onto one of several OSM map editors and duly edited. The maps are rendered and are, as stated on the website, made available to the world; or at least to those with a broadband internet connection. From the OSM wiki:

> The Mapping Party is a convivial, community event. After the mapping is finished, the participants share food and drinks, and enjoy themselves. It's a party, after all![4]

Why Witham?
Again, from the wiki:

> Although a residential land-use area has been added for Witham, very few of the roads have yet been added. Checking the 'nonames' layer also shows that most of the roads currently mapped are lacking street names. As an introduction to OSM, Witham therefore lends itself to both GPS mapping to add missing roads and paths, and walking papers to add missing road names and Points of Interest (POI). It is believed that there is also a shared use cycle/foot path running parallel to an ever increasing length of the A12 which is currently unmapped. Some of this might also be mappable.[5]

A fieldwork scrapbook (Figure 2.2):
It is an early start on a Saturday morning, travelling from Oxford to Witham via London. Changing trains at Didcot, boarding Coach D. Peering down the

| Provisional timetable | Fieldwork technologies |
|---|---|
| Coffee available from 09:30 | 1 x Garmin™ eTrex HCx GPS device |
| 10:00 An introduction to OpenStreetMap | 1 x digital camera |
| After the introduction, go mapping, or stay and use the promised wifi to demonstrate OSM uses and methods to any interested attendees | 1 x mini video-camera<br><br>1 x voice recorder to note points of interest, landmarks and street names |
| 12:00 to 12:30-ish, lunch break starts | 1 x notebook |
| 13:30-ish to 14:00-ish, head out mapping again, or stay and demonstrate uploading the collected data | 1 x laptop for uploading GPS data<br><br>Walking papers and paper base-maps for digital upload later |
| 14:00-ish, brief interlude for Mensa regional meeting | |
| 15:30-ish to 16:00 – clearing the room. Call in to say goodbye or map until you head home | |

**Figure 2.2** Fieldwork scrapbook extract from author participation in OSM mapping party (author's image).

**STOP PRESS**: The GPS device has just arrived in the post. Excellent news.
**LONG. -1° 15' 59" W; LAT. 51° 44' 53" N, heading south-easterly**

**Figure 2.3** GPS route tracing author's movement south-easterly while mapping Witham (author's image).

carriage to see a dozen or so early risers staring at newly installed television screens on the backs of seats. No one has paid the £3.95 subscription to watch re-runs of trash television. Instead everyone is making do with watching a yellow spot flashing in the centre of a map. As the train pulls out of the station (speed: 16, 17 … 18mph; altitude: 177ft), the flashing yellow spot remains fixed in the centre, but the base map slips from right to left at jagged intervals; from Didcot, down a touch to the south-east, and on to Reading (figure 2.3). In this instance, CloudMade, a company founded by OSM enthusiasts and now sponsoring OSM technically and financially, has used the base map generated by OSM (the map created and edited by OSM users) and customised it for the needs of the train company; superimposing a separate GPS-controlled

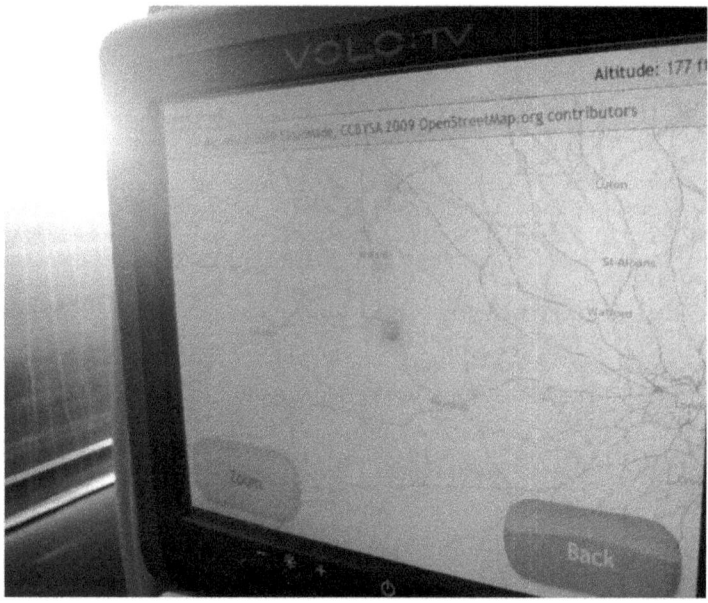

**Figure 2.4** Author's train journey photograph demonstrating the train operator's use of OSM base maps in passenger navigation screens (author's image).

cursor upon the base map to geolocate the train for the duration of the journey (Figure 2.4); a good example of a mapping-mashup using copyright restriction free data. Sitting down, rushing all too earnestly to get the map displayed on the individual back-of-seat screen; surprise and smugness on realising that the map had been generated with OSM data. Serendipity; the anorak, vernacular, geeky subjectivities of train-spotting and the lines, contours and legends mapping collide in Coach D. Crowd-source mapping was making its mark in public.

The emerging map plays into the (un)folding of both the journey's duration and the modulation of the carriage's muted affective atmospheres (Bissell, 2010). But the earlier sensation of smug amusement is an isolated one; Bergson provides the cold water:

> it may, perchance, have happened to you, when seated in a railway carriage ... to hear travellers relating to one another stories which must have been comic to them, for they laughed heartily. Had you been one of their company, you would have laughed like them; but, as you were not, you had no desire whatever to do so. (Bergson, 2008: 11)

Onwards.

**LONG. 0° 38' 11" E; LAT. 51° 47' 55" N**

**Figure 2.5** Author GPS track (author's image).

'I don't think we've met before', said Ed. No we hadn't, but he interrupted gluing posters in the window panes of the White Hart Hotel to shake hands. Ed had organised the Witham mapping party and was setting up his PowerPoint presentation that, for today only, had a dual audience as the event was timed to coincide with a regional meeting of Mensa (of which Ed is regional secretary). 'You do know there is a Mensa meeting on later this afternoon?' he asked. Yes, I knew, even if the IQ was not up to the job. Ed is from Clacton-on-Sea, not far from Witham, and is proud of his role in mapping most of Essex from Billericay to Walton-on-Naze for OSM. How did Ed get involved? His friends had been doing some mapping, so he brought himself a Navi™ GPS device and set off on his bike through the country lanes surrounding Clacton. It appealed to his professional sensibilities as a software programmer. An Oxford educated mathematician, Ed is keen on detail, and prefers accuracy to guesswork.

The previously quiet function room, carpeted in 1970s decor suddenly reverberates with the strange beats of something akin to elevator music on steroids. Ed had loaded a video called *2008: A Year of Edits*. Produced by ITO World, a data visualisation supplier which, in common with CloudMade, has a quasi-sponsoring relationship with OSM, the video shows a spinning globe set against the pulsating sound track as lines flash across parts of the planet, each flash animating a newly generated addition or edit to the OSM map during 2008. The brightest and most intense flashes occur where one might expect them to occur, in Europe and North America, but there are not many places left on the planet where the odd trace has not yet permeated. India strikes me as particularly bright. Flashing nodes of light are backed by the steady pulse of a synthesised bass drum; each pulse proclaiming OSM's mark, each beat beckoning for more to come; an expectant, anticipatory rhythm of simultaneous cartographic past-ness and futurity.

The beat ongoing, we discuss the recent efforts of OSM in mapping Haiti following the massive earthquake that struck the country in January 2010. Ed did a lot of mapping of south-west Haiti, deliberately avoiding Port-au-Prince;

the area that had received most of the mappers' attention and cartographic efforts. The difference in OSM coverage between the pre- and post-earthquake maps is stark; more so because these maps had been rendered without taking a single step in Haiti itself, but traced remotely from CIA and Yahoo satellite imagery.

Pause, for one second; perhaps that is not entirely true. We look at the 'raw' data base layer of Haiti; there are a cluster of thin blue lines in the corner; blue lines denote user-generated GPS traces, which in turn suggests that there are OSM mappers actually present in body in Haiti. I'm surprised, Ed is taken aback. Until, that is, we zoom out and realise we are not looking at Haiti at all; we are gazing at a base layer image of Clacton, Essex. Awkward, but demonstrative of a vernacular myopia when everything is reduced to its cartographic data primitives such as *nodes*, *ways* and *relations*. Clacton and Port-au-Prince do not appear that dissimilar; a pointer towards the familiar limits of abstraction from the world and the assumptions made about those abstractions. When the world is packed into lines as it is in this case, it serves as a reminder of how pervasive cartographic reason has been in Western thought and governance (and by extension, in everyday life); that disparate assemblages, cultures and things can be convened and disciplined through geometric abstraction.

OSM user 'Nigel' sends a text to Ed, he can't make it today, he is going to the theatre in Yorkshire tonight; a long drive ahead. Participation at such events is always precarious.

Ed was preparing an introductory talk on mapping techniques and editing software for his Mensa colleagues and other interested mappers, of which there were not that many, fourteen of us in total, if that. He outlines OSM's reasons for being, and the utility of the map; it can be crafted to any use the user wants, whether it be for cycle maps, maritime charts or, apparently, icing cakes. Cakes are in fact integral to OSM lore for two reasons. First, the areas to be mapped are transformed into 'cakes', or rather diagrams that are used to divide the area of concern into identifiable, digestible tracts; a cartographic mapping tool borrowed from baking to order, discipline and manage who maps what. Second, cakes form an important litmus test of whether a map is genuinely free from licensing restriction. For example, to ice a cake with the transfer of a map sourced from Google would be to violate Google Maps' Terms of Service (which states that derivative works, even iced cakes, are not permissible). Conversely, to ice a cake with a transfer of an OSM map is entirely acceptable, assuming that the resulting cake is available to share, in both its derivative forms and its devouring. It is in these micropolitical (and simultaneously macropolitical) gestures that OSM and similar groups stand out from their traditional cartographic antecedents; characterised by a stubborn refusal to be reined in by copyright restrictions or corporate vindictiveness, and instead marked by an editing of the world that produces other forms of knowing and navigating space, whereby the map is not mimetic, but instead a device, a tool, a technique, a cake for being in the world.

*2008: A Year of Edits* is stuck on a continuous loop; grating, dislodging time-spaces to think. Mensa members interject with questions; 'what scale do you render the maps?' 'Isn't that data copyrighted?' 'If we mark a waypoint in front of a pub, then we'll be marking where our bodies are located in front of the pub, not marking the pub itself; how can we be accurate?' 'What projection are we using here...a transverse Mercator, yes?' A transverse Mercator?[6] Even Ed was stumped. Data copyright? This is an awkward one. In April 2010, Ordnance Survey was forced into releasing a vast amount of geo-data that had previously been under strict licensing and commercial control. Not anymore; now anyone can access the Ordnance Survey 'Open Data Service', which includes street map data. Consequently, to reproduce this data without permission is now entirely legal. A great success surely; the outcome OSM had been fighting for ever since its inception? Technically yes, but Ed looks a little forlorn. Now that OSM users can use Ordnance Survey maps as base-level, third-party imagery, he was concerned that no one would turn up to Witham. Why get up and walk around Witham when one can sit at home and transpose data from one map to the other? Indeed, why bother at all? All the questions that emerged digitally in the mail-lists previously now find themselves posed in analogue.

Tellingly, OSM urge users to cultivate experiences in an empirical and bodily sense, to allow the 'field' to underpin and legitimate the maps generated. Experience being understood here as, 'the instant field of the present [that] is always experienced in its "pure" state, plain unqualified actuality' (James, 2003: 69); experience as the cornerstone of vernacular activity. Paul Carter asks the question, how can we live with our maps differently, away from an understanding of maps that renders them as hegemonic or disempowering (Carter, 2009)? One response might be that editing the world with OSM offers a way of living differently with our maps through the valorisation and pursuit of experiential cartographies that involve messy movement traces, 'a geo-graphy, where geography merges into performances to produce a different design in the surface' (Carter, 2009: 14); a vernacular mapping.

Bodily experiences too are central to the performances of being an accomplished amateur as Hennion (2007: 101) remarks, 'the meticulous activity of amateurs is a machinery to bring forth through contact and feel differences infinitely multiplying'. However, simply because OSM relies on the work of amateurs themselves working through and generating vernacular sensibilities, it does not follow therefore that some form of disciplining is absent from the practice of vernacular mapping. On the contrary, these vernacular impulses are trained by informal hierarchies of experience, technologies and conventions that promote a regime, or at least, a sensibility of self-governance and self-discipline. For example, in relation to disciplining the amateur, cartographic 'errors' present in the map are detected by automatic scripts, algorithms and programme robots (created by OpenStreetMappers who possess programming skills). On detection, errors are deleted automatically, a record of which is posted publically on the OSM wiki; a list-like discipline of name-and-shame that quickly striates the amateur-amateur from the amateur-expert, and to follow Schaffer (1988: 119) in relation to amateurs and astronomy, '"mere" observers [become] relegated to the base of a hierarchy of management and vigilance, inspected by their superiors with as much concern as … the stars themselves'.

> Mapping: 'It takes close attention to see what is happening in front of you. It takes work, pious effort, to see what you are looking at'. (DeLillo, 2010: 16)

What might count as the cartographic node of vernacular or micropolitical sensibility here is the manner in which the OSM mappers attend to their matters of concern in a 'perplexed' mode. To paraphrase Hennion once again, for OSM mappers to work in perplexed mode means to be on the lookout for what mapping does to them, to learn to be affected, to be 'attentive to traces of what [mapping] does to others; a sharing out among the direct sensations to be experienced (or whose experience is being sought), and the indirect relays that permit one to change one's own judgement a bit, while relying in part on the advice of others' (Hennion, 2007: 104). OSM could be said to rely upon these embodied experiences as central to its propagation and maintenance because it is in these experiences that a particular sociability and micropolitical sociability is cultivated; sensibilities that sustain and promote a variegated, creative and sometimes baffling expertise; the hallmark of OSM's cartographic energies. The cartographer does not precede the mapping. Instead the cartographer and the mapping are co-constitutive, always in the making.

Circle back to the event. Witham, still.

Ed has finished his presentation, and spare GPS devices are loaned to some of the other first-time mappers. Everyone starts pushing buttons. Some walk out of the pub straight away, they know what they are doing. They are OSM veterans. For others, looks of extreme bewilderment become etched on their faces. It is going to be a long morning. 'Don't worry', says Ed. 'OSM is just about *nodes*, *ways* and *relations*'.

Stop there. Nodes, ways, relations?

**Node**: A *node* is the basic element, building block, of the OSM scheme. Nodes consist of latitude and longitude (a single geospatial point).
**Way**: A *way* is an ordered interconnection of at least two and at most two-thousand nodes that describe a linear feature such as a street, footpath, railway line, river, fence, power line, area or building outline. One way is characterised with uniform properties; for example, priority (motorway, trunk roads), surface quality, speed, etcetera.
**Relation**: A *relation* is a group of zero or more primitives (nodes and ways) with associated roles. It is used for specifying relationships between objects, and may also model an abstract object.

Nodes, ways, relations; the OSM data 'primitives', the cartographic building blocks of the OSM map, but taken further they are also constitutive of a geometry of temporalities and subjectivities in the making. If nodes, ways and relations are the digital protocols of how the map can be drawn and edited, then so too are they the analogue and affective functions of a vernacular subjectivity; cartographers as nodes, as individuated actants, but nodes only insofar as they are put into motion with other mappers and technologies to generate ways and relations; ways as technologies bringing cartographies into being, and relations as the necessary co-fabrication of both the maps themselves, but also the space-times and events in which the map is forged; 'the relation itself is a part of pure experience; one if its "terms" becomes the subject or bearer of the knowledge, the knower, the other becomes the object unknown' (James, 2003: 4). Micropolitical, vernacular subjectivities are thoroughly relational; co-constitutive relations and assemblages of technicity, expert and amateur bodies. Such subjectivities promote creativity and expertise outside of and beyond proscribed institutional norms. Much like the movement embedded in maps, to paraphrase Carter, this geometry of subjectivities is neither fixed nor linear, but instead rhythmic; a pulsating geometry of relations and creativity through which maps and micropolitical gestures emanate (Carter, 2009). Nodes, ways and relations, then, are more than editorial metaphors for the assemblages at work here, instead they are more urgently caught up in the very cultivation of these cartographic practices and spatio-temporal assemblages.

Nodes, ways, relations, then.

Still some looks of bewilderment and the pace of the day thus far has been sluggish at best. Time to inject some impetus. 'Let's just get out there', suggests Ed; good idea. First, we need to cut the cake, or more specifically, the sector cake of Witham; the town divided into palatable areal chunks that lack roads, points of interest, or are indeed devoid of mapped space altogether. We pick

**LONG. 0° 38' 11" E; LAT. 51° 47' 52" N**

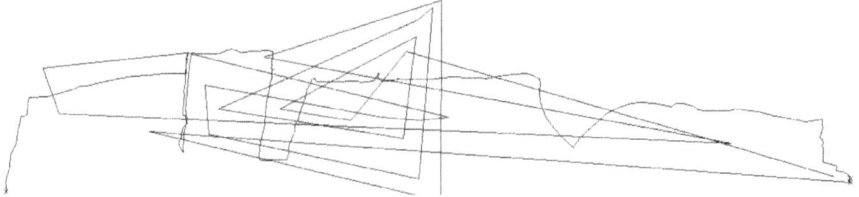

**Figure 2.6** Author GPS track (author's image).

the slices of Witham that we want to traverse and map (see Figure 2.6). Some pick slices nearby the pub, some pick slices that involve a riverside walk. I pick slice eighteen. No more than a lucky number, a gut feeling, and something alluringly trapezoidal about the slice. Ed passes me a walking paper for slice eighteen on which he has already kindly outlined the sector with an orange highlighter. Sipping quickly the last dregs of coffee, we say farewell for now. Slice eighteen is ten minutes' walk south of the pub.

## Acquiring satellites

So this is it. Isolation.

That, as ever, is not strictly true; the GPS, the voice recorder, the walking paper, the camera, the fieldwork mentality; faithful companions all the way.

On reaching the edge of slice eighteen I am immediately lost. The edge of the cake starts on Laurence Avenue, but I'm stood on Howbridge Road. Looking at the walking paper, and other maps of Witham I had printed the day before, an irony grows; that to make a map is to need a map of the area you want to map. Standing under the nearest bus stop for shade, searching desperately for Howbridge Road to transform into Laurence Avenue, but the map is having none of it. Orwell's (1938: 58) Spanish civil war encounter with a naive militiaman in Catalonia springs to mind, 'obviously he could not make head or tail of the map; obviously he regarded map-reading as a stupendous intellectual feat'.

The mapping stops before it even begins.

'What good is time?', Henri Bergson (1992: 93) asks mischievously. Not much good, if we are to take time to be a matter of chronological measure. Consider, for example, how the history of cartography is problematic insofar as it can be told as a linear, hubristic tale of modernisation and progress. Vernacular mapping is not reliant upon such metrics of the clock and the timetable. It has no time for such time. Bergson provides us with another formulation of and for time, premised on the metaphysical acceptance of indivisibility and the reality of constant change; 'we shall think of all change, all movement, as being absolutely indivisible' (Bergson, 1992: 142). This, unsurprisingly, extends to Bergson's figuring of time as 'duration'. While Bergson admits that there is 'succession' in the movement of time, he refuses the neat compartmentalisation of durational time into chronological units. To be adumbrating about duration, it is to state that past, present and future are constitutive of one another – they enfold, they are indeed, indivisible; duration cannot be measured, it can only be intuited, and, as Deleuze (2004: 33) proclaims, 'intuition is the joy of difference'. Perhaps in the annals of cartography, a mapper's reliance on sensibility, intuition and virtuality is underplayed at the expense of realising the unabated difference that maps can produce. And don't think of intuition, *intuitively*; to intuit takes skill and requires a kind of intellectual and somatic empathy with the matter that one is interested in; an 'intellectual sympathy by which one places oneself within an object in order to coincide with what is unique in it and consequently inexpressible' (Bergson, 1992: 23). That duration is not chronologically metered does not mean that there are no contours, disruptions or undulations in duration; crucially, duration still permits an acknowledgement of rhythm in producing space and life itself.

Mapping is replete with concatenated rhythms that call upon the durational movement of pastness, presentness and futurity. Take, for example, the following encounter between Earth and its immediate orbital space.

Lost, still. Turn on the GPS device, this will help for sure. 'Acquiring satellites' reads across the GPS screen. 'Acquiring satellites' – when did that become an everyday matter? Such acquisitions are now minor moments of geopolitical connection that have become a necessary step for locating position in the absence of accurate paper maps, local knowledge and common sense. After two long minutes, the GPS has a three-dimensional lock on my position, meaning it can detect both altitude and geodesic position. Cosmic numbers to be sure. Back down to Earth, quickly, though suffusing the duration of mapping is a continuous spatio-temporal 'lag'. A signal's journey between satellite and handheld device might appear quick from a cognitive, or even straightforwardly chronological, perspective, but there is still a delay as signals negotiate with non-human interfaces before travelling the leagues between Earth and orbital space. This spatio-temporal lag is liable to create a 'slowing down' of events, but from a durational perspective, it protends the present (extending a present-ness, or 'now' by stretching the past and future). Thus, as the signals reach the handheld device from a contracted 'past', the waiting, anxiety and anticipation for signal recognition opens up a virtual space-time in which mappers can learn to map differently, or to be 'creative' in the way they manage that lag. Lag, insofar as it is evidence of a durational, temporal 'stretch', necessitates creative, vernacular responses.

The lag extends/protends, and disorientation grows.

A road name would do for now, but even that is not forthcoming, as the GPS is only pre-loaded with a perfunctory road map of the world. A further £150 would buy a premium upload for UK road data (Ed later tells me that I should have downloaded OSM base maps on to the GPS for free. I'll do that when I get home). The GPS is not helping, so in walking a little further to the next street, I realise that I am in the right place, but that the OSM walking paper is inaccurate; not entirely unexpected, as this was one of the very reasons for convening a mapping party in the first place. My first edit; *delete* Laurence Avenue, *replace* with Howbridge Road; just scribbles on the walking paper. Stood on the corner of Pelly Avenue; this is where I'll make a start. It does not yet exist on the OSM map, so for all concerned, it is somewhat new territory. There was no such novelty for the passer-by who asked if I was lost; 'are you ok, you look lost?'. Spot on. I have no idea where I am, but I explain what I'm doing, and thank them for their help. They walk off, nonplussed. The GPS device is back on; holding down firmly the cursor button on the GPS to mark a waypoint. Point 001. Over the course of the day, I mark seventy-two waypoints, designating street corners, points of interest, and anything else that struck me, literally on occasion, along the way.

Mapping, vernacular or otherwise, demands multi-tasking; sketching an outline on the walking paper as one goes, maintaining orientation, marking waypoints, checking the status of the GPS, labelling all points of interest (including postboxes), taking photographs, making voice recordings, ensuring to walk down every road, every pathway, every cycle lane, making sure that the trajectory of one's walking route is not too erratic so as to generate a bizarrely stochastic electronic breadcrumb trail on the GPS which would become a nightmare to upload and render into a map; yes, the rhythm can be as breathless as it reads. While the GPS takes care of the lines walked, I struggle to mark points of interest in-between the lines, approximating their position on the paper map. At the same time, I and the other mappers (wherever they have gone) need to appear inconspicuous and non-intrusive. Understandably a few glares from residents emerged, watching my cyclical and repetitious movements around the estate and forced at one point into a double pirouette to calibrate the GPS compass. Irksome, I already had a sense of where north was without the dance, but at least now we're calibrated. More importantly, the GPS acts as an authenticator of presence, a reliable witness in the mapping; it feels that to be without one would be to jeopardise the sense of cartographic legitimacy OSM is trying to engender.

Walking with the head down, peering at the traces emerging on the GPS, pacing at around 3.7 miles per hour (is that slow?), but there is no sense of scale, so pencil lines are crammed, depicting roads into an unnecessarily small sketch map, not entirely confident that this is not a mistake, or that experienced OSM users won't make derisory comments about the day's mapping when GPS traces are uploaded on to the OSM server. The frenetic rhythms of charting, anticipating and walking have a disparate effect on cartographic attention. On the one hand, the stuttering rhythms provoke simple lapses in concentration; what was that street called, where have I just been? On the other hand, the durational quality of the mapping conversely focuses a different kind of affective attention to the virtual; the on-the-cusp anticipation of what's next, where am I headed? To this end, and from an instrumental perspective, one might argue that given the idea of cartographic duration, neither a slowing down or speeding up of the mapping's rhythm would necessarily enhance or worsen the quality of the cartography. Back to space, and from a God's trick perspective, the number eighteen slice doesn't appear that large. But of course walking in and out of its labyrinthine roads and pathways soon elucidates the feeling of considerable areal space, and a realisation that a cartographer will never capture everything that is here, even in a lifetime of mapping. This was going to take several hours, so I ring Ed and tell him I'm skipping lunch to continue mapping (it extricates me from the Mensa meeting too).

Relations with the GPS device are ambivalent. On the one hand, their helpful triangulation of position can provide moments of endorsement; guarantors of location. 'You are here'; you are in the right place; human and non-human in geometric harmony. On the other hand, they stir tension, argument and tears, setting bodies off on labyrinthine nightmares, to the edge of their worlds. Safety net or banana skin, GPS devices have replaced protractors and compasses as the cartographer's instrument of choice; powerful remote-sensors that have revolutionised mapping procedures, but devices that have something of the Latourian black-box about them. As suggested earlier, calibration failures, glitches, and even lags, far from being setbacks, are generative points of contention, talking points for cartographic hobbyists; problems that require solutions. In sum, despite OSM's strive for accuracy, the ethos of failure is one that is cultivated and celebrated as a productive ethic in experimentation; if you have an idea, or a glitch, do not, as OSM's founder suggests, worry. Instead, 'JFFI: Just F*cking Fix It' (Coast, 2010: no pagination). This is, after all, vernacular.

Continue walking, continue mapping slice eighteen. A Witham's summer day, the neck starting to feel the sun's presence more urgently; a somatic, durational beckoning for the day to be done – soon. There's Howbridge Junior School, there's the Jack and Jenny Pub, there's the bus stop for service ninety.

**LONG. 0° 38' 11" E; LAT. 51° 47' 52" N**

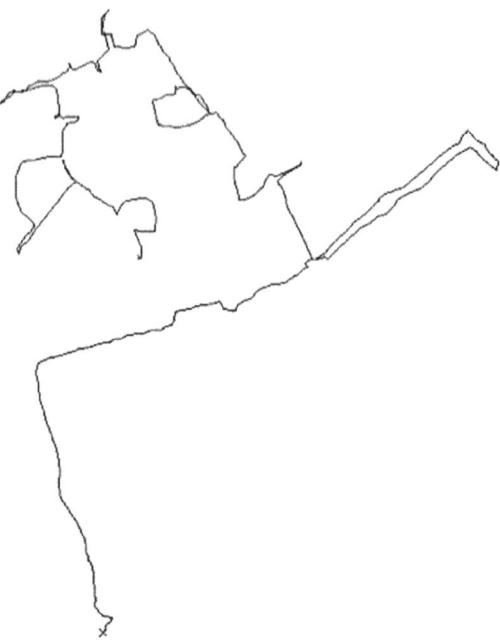

**Figure 2.7** Author GPS track (author's image).

All waypoints, all useful nodes of information; mining for data, gorging on GPS signals; guilty pleasures for sure. Tongue in cheek, this feels like a civic service, a good deed in the mapping, though this kind of vernacular mapping shouldn't be beholden to such stilted, do-gooder politics. Slice eighteen has a rhizomatic quality about it; there is a road through the gap in the houses, but I don't know how to get there, and I cannot be sure that the path has not already been walked. Two, slightly off-parallel thick lines on the GPS suggest that indeed I've been here before, so doubling back on myself, a pathway missed has been found. Kids on bikes ask me what is going on, so I ask for their help; which streets do they live on, what cycle tracks do they use? It is a brief but useful vernacular cartographic event, a mini-convening of publics; a distributed kind of wayfinding with others and a type of conversational mapping which has a heritage far longer than paper bound, Euclidean cartography.

In duration, four hours pass extremely quickly, but in a spirit of cautiousness, I loop around slice eighteen once more, treble-checking that the GPS traces have been saved to the memory card, with waypoints intact. Time to head back to the pub. A few other OSM users had turned up who had not been in attendance in the morning. Most of them drove their cars around the bypass and environs of Witham, collecting vast amounts of geo-data from just a few hours work. Our function room booking is about to expire, chronological time is against us, so Ed and the others provide a rapid tutorial on uploading the data to the OSM server and then using various OSM editing packages to render the map. One last question before we depart: 'Do you find OSM mapping addictive?', I ask the assembled mappers. 'Addictive is probably the wrong adjective to use', replies Ed, offering no alternative. This is the problem when academics attempt to impose the categorical upon the non-categorical. So what is this vernacular mapping? Perhaps it is an obsessive compulsion in the mapping, the quest for spatial affirmation and a sheer junketing for the ludic qualities of this kind of mapping (Perkins, 2009); perhaps it is an experiential *jouissance* to mapping not found in the state-led cartographies of early nation-state Europe. Maybe vernacular mapping, and part of its contemporary appeal, is about the opportunities for time-travel, in durational sense. It could be all, or none of the above. In one sense, the likes of OSM dos not mark reclamation of mapping for the everyday, as people are mapping all the time in some form or other, but what it does do is disturb and inflect the meanings of Euclidean cartography; where what counts as accurate and precise is up for grabs.

## Coda

> I tried not to think beyond geography. (DeLillo, 2010: 101)

Those urges to capture everything, to mark nodes, ways and relations, to monopolise movement cartographically, to laud diagrammatically over Witham as Captain James Cook did so in Botany Bay, Australia (Carter, 1987); where do these urges come from? In part, they stem from the discursive and cultural baggage that has become welded to cartography; the discipline's associations with colonialism, monopoly and meta-narrative. Importantly, the focus of this chapter on the experiential is not meant as a disavowal of other well-known linguistic and discursive registers of cartographic practice. Following Wylie (2002), to divorce the experiential from the discursive in its entirety would be somewhat disingenuous. In the case of pacing around Witham, swashes of Cartesian and Euclidean thinking pervade the somatic rhythms of mapping, as bodies are simultaneously orientated and disorientated by cartographic norms and orthodoxies that have been codified and handed down through generations, through textbooks and now through wikis.

That does not then entail a valorisation of Euclidean cartography or Cartesian logic, it is merely a way of writing-otherwise these mappings, of thinking through how notions such as objectivity and drawings such as straight lines, while seemingly disciplinary and stifling, can actually be generative of a panoply of bodily dispositions, and micropolitical possibilities. Put simply, to achieve and trace a straight line for OSM in Witham, bodies must move, sweat, ache, meander and blister. They need to coalesce with non-human bodies and learn to be affected in the process. It is then the virtual potential of these cartographic bodies in motion that invigorates these kinds of cartographies, whereby their production and affects are recombinant and continually at risk, not pre-figured as per the cartographies of past officialdom.

What of the time travelling engendered by these moments of participatory cartography? This requires a return, obliquely, to Henri Bergson's (1992) thinking on time. If, as Bergson implores, time is figured qualitatively as 'duration', rather than as 'chronological', then the former makes indivisible the fractures between past, present and future and transforms, instead, their matter to exist on the same plane of immanence. To read time by the latter, chronologically, for Bergson, is to spatialise time – to crystallise spatial units of time into the hours, the minutes, and the seconds that flow by. In other words, Bergson is castigating a cartographic figuring of time and so on this claim alone, a map would be understood as freeze-frame of a particular time. Yet, not only is Bergson's rendering of space ironically reductive, but participatory mapping also helps to illuminate the indivisibility of time and space, and indeed to demonstrate the durational qualities of space (and not just of time). Mapping here is not about creating spatially or temporally bound segments of experience in order to create similarly segmental abstractions. Instead, nodes, ways and relations are interstitial points of departure; or to borrow from James Ash (2015), they might be figured as 'envelopes' of space-time; foldings that call on the past, that beckon a futurity while simultaneously propagating a cartographic present. The temporally awkward narration of these mapping vignettes points to the indeterminacy of the future, even if much of the mapping is seemingly mimetic, based upon already well-annotated streets and pre-figured way points. This might not be time-travel in a register figured by H. G. Wells, for, as Elizabeth Grosz (2005: 10) states, 'durational force, the force of temporality is the movement of complication, dispersion or difference that makes any becoming possible and the world a site of endless and unchartable becomings'. Quite so. Why fix the world cartographically in the chronological past, lest difference be numbed? Nonetheless, I would argue that becomings are 'chartable', but only if one charges that charting, or that cartography with a vernacular ethos which itself hinges on never-quite-knowing; of finding joy in intuition and wayfinding through spatial and temporal disorientation. Cartographic temporality then, in a vernacular sense, and as exemplified to a degree by

OSM, might be understood as a tentative, speculative time-travel; a mapping not based on the certitude of chronology, but through the spatial-temporal indivisibility of duration.

## Notes

1  Boria (2013) also notes the increasingly strained relations between geographers and cartographers.
2  The 2010 eruptions of the Eyjafjallajökull volcano in Iceland caused widespread disruption to aviation across north-west Europe for several weeks. The matter of concern or manifested controversy here, however, was less the particles of volcanic ash themselves, but their cartographic rendering and modelling. Competing measurements of aircraft tolerance to differing levels of ash by multiple jurisdictions highlighted the awkward, mischievous and politically febrile entanglements of scientific representations (see Ulfarsson and Unger, 2011).
3  OpenStreetMappers are referred to by their chosen user names, e.g. TimSC and SteveC.
4  Mapping parties. [Online] Available at: http://wiki.openstreetmap.org/wiki/Mapping_parties (accessed 23 November 2017).
5  Witham/201004 Mapping Party. [Online] Available at: http://wiki.openstreetmap.org/wiki/Witham/201004_Mapping_Party (accessed 23 November 2017).
6  Developed protractedly by Carl Gauss and Johann Krüger in 1825 and 1912 respectively, the Transverse Mercator projection retains the central meridian line of the original Mercator projection, but unfolds the image of the Earth in a concertina fashion so as to display the terrestrial poles (Jackson, 1978) and narrow the proportional areal disparities between landmasses.

## References

Ash, J. (2015) *The Interface Envelope: Gaming, Technology, Power*. London: Bloomsbury.
Bergson, H. (1992) *The Creative Mind: An Introduction to Metaphysics*. New York: Carol Publishing.
Bergson, H. (2008) *Laughter: An Essay on the Meaning of the Comic*. Rockville, Maryland: Arc Manor.
Bissell, D. (2010) Passenger mobilities: Affective atmospheres and the sociality of public transport. *Environment and Planning D: Society and Space*, 28: pp. 270–289.
Boria, E. (2013) Geographers and maps: A relationship in crisis. *L'Espace Politique*, 21. [Online] Available at: http://espacepolitique.revues.org/2802 (accessed 3 August 2016).
Burns, R. (2014) Moments of closure in the knowledge politics of digital humanitarianism. *Geoforum*, 53: pp. 51–62.

Caquard, S. (2014) Cartography II: Collective cartographies in the social media era. *Progress in Human Geography*, 38: pp. 141–150.

Carter, P. (1987) *The Road to Botany Bay: An Exploration of Landscape and History*. Minneapolis, Minnesota: University of Minnesota Press.

Carter, P. (2009) *Dark Writing: Geography, Performance, Design*. Honolulu: University of Hawaii Press.

Chambers, R. (2006) Participatory mapping and Geographic Information Systems: Whose map? Who is empowered and who disempowered? Who gains and who loses? *Electronic Journal of Information Systems in Developing Countries*, 25: pp. 1–11.

Coast, S. (2010) State of OpenStreetMap keynote address. *State of the Map Conference*. Girona, Spain. [Online] Available at: http://wiki.openstreetmap.org/wiki/State_Of_The_Map_2010 (accessed 3 August 2016).

Crampton, J. (2010) *Mapping: A Critical Introduction to Cartography and GIS*. Chichester: Wiley-Blackwell.

de Certeau, M. (1984) *The Practice of Everyday Life*. Berkeley, California: University of California Press.

Deleuze, G. (1995) *Negotiations: 1972–1990*. New York: Columbia University Press.

Deleuze, G. (2004) *Desert Islands and Other Texts (1953–1974)*. Los Angeles: Semiotext(e).

DeLillo, D. (2010) *Point Omega*. London: Picador.

Elwood, S. and Leszczynski, A. (2013) New spatial media, new knowledge politics. *Transactions of the Institute of British Geographers*, 38: pp. 544–559.

Gerlach, J. (2014) Lines, contours, legends: Coordinates for vernacular mapping. *Progress in Human Geography*, 38: pp. 22–39.

Grosz, E. (2005) Bergson, Deleuze and the becoming of unbecoming. *Parallax*, 11: pp. 4–13.

Hennion, A. (2007) Those things that hold us together: Taste and sociology. *Cultural Sociology*, 1: pp. 97–114.

Jackson, J. (1978) Transverse Mercator projection. *Survey Review*. 24: pp. 278–285.

James, W. (2003) *Essays in Radical Empiricism*. New York: Dover.

Jellis, T. (2015) Spatial experiments: Art, geography, pedagogy. Cultural geographies in practice. *Cultural Geographies*, 22: pp. 369–374.

Kitchin, R. and Dodge, M. (2007) Re-thinking maps. *Progress in Human Geography*, 31: pp. 1–14.

Lacoste, Y. (1973) An illustration of geographical warfare. *Antipode*, 5: pp. 1–13.

Latour, B. (1993) *We Have Never Been Modern*. Cambridge, Massachusetts: Harvard University Press.

Latour, B. (2008) Review essay: The Netz-Works of Greek deductions. *Social Studies of Science*, 38: pp. 441–459.

Leszczynski, A. (2014) On the neo in neogeography. *Annals of the Association of American Geographers*, 104: pp. 60–79.

Lorimer, H. and Wylie, J. (2010) LOOP (a geography). *Performance Research: A Journal of the Performing Arts*, 15: pp. 6–13.

McCormack, D. (2014) Atmospheric things and circumstantial excursions. *Cultural Geographies*, 21: pp. 605–625.

Newling, J. (2005) *Writings by John Newling 1995–2005*. London: SWPA.
Orwell, G. (1938) *Homage to Catalonia*. London: Penguin.
Perkins, C. (2009) 'Playing with maps'. In: Dodge, M., Kitchin, R. and Perkins, C. (eds) *Rethinking Maps*. London: Routledge, pp. 167–188.
Perkins, C. (2014) Plotting practices and politics: (Im)mutable narratives in OpenStreetMap. *Transactions of the Institute of British Geographers*, 39: pp. 304–317.
Perkins, C. and Dodge, M. (2008) The potential of user generated cartography: A case study of the OpenStreetMap project and Mapchester mapping party. *North West Geography*, 8: pp. 19–32.
Rorty, R. (2009) *Philosophy and the Mirror of Nature*. Princeton, New Jersey: Princeton University Press.
Schaffer, S. (1988) Astronomers mark time: Discipline and the personal equation. *Science in Context*, 2: pp. 115–145.
Stengers, I. (2005) Introductory notes on an ecology of practices. *Cultural Studies Review*, 11: pp. 183–196.
Thrift, N. (2012) The insubstantial pageant: Producing an untoward land. *Cultural Geographies*, 19: pp. 141–168.
Ulfarsson, G. and Unger, E. (2011) Impacts and responses of Icelandic aviation to the 2010 Eyjafjallajökull volcanic eruption: Case study. *Transportation Research Record: Journal of the Transportation Research Board*, 2214: pp. 144–151.
Whatmore, S. (2002) *Hybrid Geographies: Natures, Cultures, Spaces*. London: Routledge.
Wylie, J. (2002) An essay on ascending Glastonbury Tor. *Geoforum*, 33: pp. 441–454.
Zook, M., Dodge, M., Aoyama, Y. and Townsend, A. (2004) 'New digital geographies: Information, communication, and place'. In: Brunn, S., Cutter, S., Harrington, J. (eds) *Geography and Technology*. Dordrecht: Kluwer Academic Publishers, pp. 155–176.
Zook, M. and Graham, M. (2007) The creative reconstruction of the internet: Google and the privatization of cyberspace and DigiPlace. *Geoforum*, 38: pp. 1322–1343.

# 3
## Mapping the quixotic volatility of smellscapes: a trialogue

*Sybille Lammes, Kate McLean and Chris Perkins*

The following trialogue recounts a three-way conversation/interview. During our exchange words and ideas overlapped, occurring simultaneously, or slightly offset, to two or all three of us at any one moment in time. The subject of our conversation was the temporality invoked within the practices of smellscape mapping. To recount the conversation as sequential would be disingenuous, omitting any moments of cross-thinking, as well as leaving the visuals, which were integral to the conversation, to one side.

Instead we use a page layout inspired by the *Chronicles of Eusebius*. A first in layout design in the fifteenth century as the codex started to replace scrolls, these printed Chronicles showed a comparison of historical data with synchronous events depicted in tables for the first time. Eusebius' aim was to establish the place of Christianity and also synchronise the chronologies of the historical narratives of several nations. His design used columns to transliterate between languages:

> Nineteen parallel columns, one to a nation, traced the rise and fall of the ancient Assyrians, Egyptians, and Persians, as well as the Greeks and Romans, who still ruled the world. (Rosenberg, 2012: 26)

Our layout design for this chapter reflects the nature of a trialogue using Eusebius' tabular system to convey the informality and the numerous interruptions of thoughts, spoken words and visuals. We invite you to join us …

**Figure 3.1** Smellmap: Edinburgh © 2010 Kate McLean. This figure has not been made available under a CC licence. Permission to reproduce it must be sought from the copyright holder.

## Introduction

Kate McLean's smell mapping practice (see Table 3.1) began in Edinburgh in 2010 with exhibited maps compiled as part of her Masters degree at the Edinburgh College of Art. The visual style of these maps and their fixed format remained largely unchanged for two years with mapping compiled for cities including Edinburgh (Figure 3.1), Glasgow, New York, Paris and Newport (Rhode Island). With the publication of 'Smellmap: Amsterdam' in 2013, and during her PhD research, McLean began to experiment with more dynamic mapping forms, including animations, a trend continued in the publication of 'Smellmap: Pamplona' in 2014. A central concern of her work is increasingly how to mediate the volatile urban smellscape in a mapping format that reflects human subjectivity, the ephemerality of smell and the uniqueness of particular smellscapes. The unseen methodologies behind her practice have also changed over this period, with an increasing incorporation of multiple voices into the published mapping, and a changing deployment of technologies.

**Table 3.1** Published 'Smellmaps' by Kate McLean

| Year | Smell map city location | Medium | Temporality | Participants |
|---|---|---|---|---|
| 2010 | Paris | Smells + hand illustration | Individual memory/association | 10 |
| 2011 | Edinburgh | Digital print + natural smells | Snapshot + personal association | 10 |
| 2011 | New York's smelliest block | Digital print | Auto-ethnographic snapshot | 2 |
| 2012 | Newport | Digital print + natural smells | Collective snapshot + personal association | 30 |
| 2012 | Glasgow | Digital print + smells | Collective snapshot + personal association | 20 |
| 2013 | Milan | 3D hand-illustrated buildings + smells (collaboration with Olivia Alice) | Personal association | 2 |
| 2013 | New York's thresholds | Digital print | Nostalgia/snapshot | 20 |
| 2013 | Amsterdam | Digital print + synthetic smells + motion graphic | Collective snapshot + personal association + ephemerality of smell | 44 |
| 2014 | Canterbury | Hand illustration + diffused natural smells | Personal association | 5 |
| 2014 | Pamplona | Motion graphic | Individual perception of smell duration/ephemerality | 58 |

| Year | Smell map city location | Medium | Temporality | Participants |
|---|---|---|---|---|
| 2015 | Singapore | Digital print + (on-going project) | Individual perception of smell duration & comparative day/night maps to be created + rhythm of urban living/futurity/ dimensionality/ ephemerality | 104 |

Further information about Kate's work can be found at http://sensorymaps.blogspot.co.uk and http://sensorymaps.com.

The conversation reported in this chapter took place on the Island of Gozo in April 2015, when Kate participated in rural smell mapping of the island, as part of the other two authors' Erasmus-funded fieldcourse investigating mobile methodologies in field-based learning – the 'Go Gozo' project. References that we made to literature appear in a bibliography at the end, and limited editing of the transcript has also taken place, to clarify issues. The temporalities that emerged during this discussion, however, remain largely as recorded in the transcript.

## 'Smell? What do you mean smell?'

*Sybille:*

*So how do you map smell, really?*

Kate:

**Chris:**

There are various approaches to the mapping of smell, from technological sampling and recreation in order to represent a moment in time, to depictions of the smellscape from the lived perspective. My approach is through looking at how people understand olfactory space, how they perceive and understand their city's smellscape. In my art practice, what I've done so far is to collect individual perceptions of urban smells in the form of 'smell notes', separating perception (information) into items of data. This process occurs as we walk along.

Yes, my practice is very subjective because smell perception in its own right is very subjective. Mapping frequently uses visual modalities as a starting point; we map according to physical features, geographic landmarks, and buildings that we see. But increasingly artists and designers are mapping the auditory elements in our landscapes (LaBelle, 2015), as well as the more ephemeral feelings such as emotions (Nold, 2009) and everyday poetic activities (Ng-Chan, 2016).

**So would you say your practice is very subjective and that is different from using other sensory modalities to map things?**

My practice differs from others in that I explore what happens when we map according to what our noses tell us.

*Could you tell us how that plays out over time? Is it a process of mapping smells through identification and selection within a certain time frame?*

Yes, the extended process involves the general public identifying smell perceptions over a period of up to ten days through 'smell walking' and then of me as the artist/designer selecting from a range of up to 2,000 smell perceptions (each of which has seven subcategories) before mapping the data manually.

It is really difficult to actually explain it to the uninitiated. Let's start from something like this … a smell walker in the process of collecting smells in a New York neighbourhood (see http://sensorymaps.com/portfolio/smellsketch-brooklyn).

The smell walker is temporarily frozen for a moment, sniffing. Prior to that instant they have been walking slowly around a city, absorbing and ingesting smells, remarking on them, deciding what to smell and where, deciding whether or not to record their 'smellervations'. When that person remarks on certain smells that they detect I ask them to write down these perceptions on paper, analysing the smell experience by recording individual data points such as the 'smell name/description'. Ten seconds later another smell walker may come along, sniff the identical bin and detect something completely different. There have been times when two people smelling inside the same bin at an identical moment in time will identify completely different smells, or nothing at all. During this particular walk in September 2014 two individuals simultaneously sniffed another bin; one noted no identifiable or perceptible odour, and the other picked out a minty whiff from chewing gum.

So the idea of a smellscape comprised of individual smells is one of fluctuation, of change, as smells evaporate and dissipate. Add to this the genetic difference of smell perception documented by Makin (2013) and we find that olfactory perception is highly nuanced and highly personal. There are methods of recording the smellscape more quantitatively, using headspace technologies, whereas mine is a very qualitative method.

## Ephemerality

*Sybille:*  Kate:  **Chris:**

*How would you relate subjectivity to temporality in your practice? I can see several processes going on: you observing and directing people, the individual and subjective rhythms of identifying smells, pausing, moving and the cumulative process of others adding subjective layers to that smellscape, sometimes in accordance and sometimes at odds. And then there is of course the fluidity of smell itself, which you have yourself*

Ephemerality can be witnessed when you try to 'catch smells' during a smell walk; in the process of recording them some simply disappear before you can go back to verify their existence. My understanding of a smellscape is as a floating visibility cloak. Yi Fu Tuan (1977: 11) says that smells are *previously called the ephemerality of smellscapes (Quercia et al., 2015; McLean, 2017).* not objects, nor can they be used to describe objects, but rather that they 'lend character to objects, rendering them visible, easy to identify and memorise'. A smellscape shifts constantly in the wind; it evaporates and dissipates as odour molecules diffuse into the atmosphere; human beings perceive individually and over time, adding to each other's subjective layers. There are multiple factors in the evanescent smellscape and this is what I'm mapping in my practice. I'm exploring a myriad of smelly temporalities.

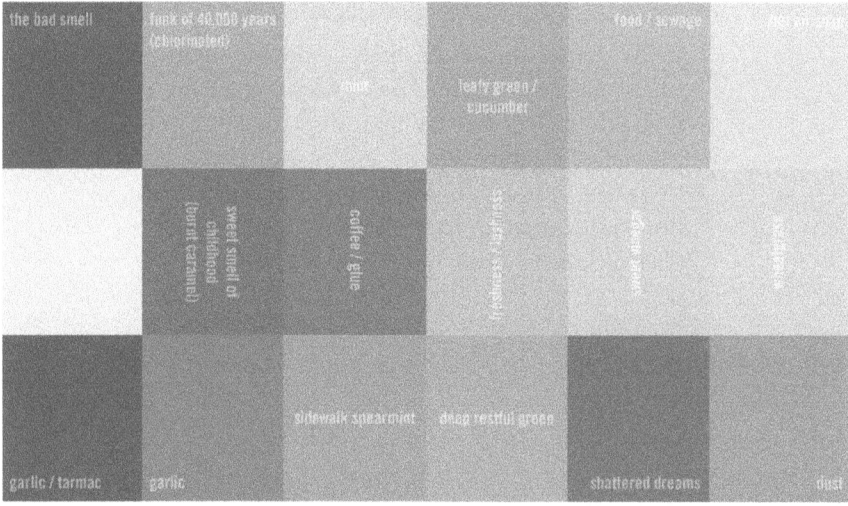

**Figure 3.2** Smellcolour sketch: Brooklyn © 2014 Kate McLean. This figure has not been made available under a CC licence. Permission to reproduce it must be sought from the copyright holder.

Depicted above (Figure 3.2) is a collated smellscape taken from about an hour and a half walk with twenty people. Each of these squares contains a single smell reference from the group where participants compared individual perceptions to agree on a smell representation at specific points on the walk.

*The duration of the walk adds another temporal axis to what's happening at this stage of creating a smellscape. And within this time frame smellers both move and stand still. So I guess as David Bissell (Bissell and Fuller, 2011) suggests, immobility is important to the temporal experience as well as mobility.*

Yes. This is a New York block. There is a spatiality, although not a literal geolocated spatiality.

**Yes, there's a spatiality to that Figure as well in the sense of someone who's walking along. This spatiality seems to emerge here as a shared consensus of a changing smellscape.**

## Temporal stories and capturing time

*Sybille:*  Kate:  **Chris:**

*We can't conceive of time without thinking of space, and especially not when mapping smells while exploring an environment. It seems to me this connection is central to your work, as you write yourself (Quercia et al., 2015; McLean, 2017). So*

**Figure 3.3** Smellmap: Pamplona (still image from movie) © 2014 Kate McLean. This figure has not been made available under a CC licence. Permission to reproduce it must be sought from the copyright holder.

*is it a spatial story, in the way that de Certeau (1984) describes, with a temporal structure that evolves during exploration?*

Yes: that's a spatial story. But then ... here's one that is both a spatial and temporal (see Figure 3.3).

*Yes, it is a matter of degree, but in the case of the Pamplona temporal story a spatial logic evolves through a temporal experience of smell. So that is a different process from creating a spatial story, where spatial touring gives rise to a narrative as a kind of temporal structure. That seems to me to be a rather important difference for understanding your work. After the walk, when you sit down and take all the participants' notes to construct a smell map, you seem to take this further by using the temporal stories of participants to create a kind of temporal map.*

**Yes, in the animation the dots are appearing ... as the dot appears, that is when you smell?**

Yeah. The dots appearing on the map are new smells manifesting in the city. Each smell has a range and intensity, as indicated by the number of concentric circles. Individually perceived smells grow, one at a time in a section of the city. The temporality of the smell shows as the shape appears to be manipulated and blown by the wind over a specific period of time which is individual to each smell.

*That beautifully captures the ephemeralities of smell, although that may seem to be a paradox, because how can you capture what is really fleeting? It really depicts the movements and depth of smells.*

**Okay, so take us through some of the factors that influence that kind of temporality then. Wind? What else?**

Wind influences longevity, even the lightest breeze will whisk odour molecules into another space. The intensity of the smell itself influences the impression of its length; some smells can be detected in tiny concentrations (such as mercaptan added to natural gas), and some scents are created or dispersed in deliberately targeted concentrations (like perfumes dispersed in retail environments). The duration of an odour is the time an individual actually perceives it lasting, from first sniffing to final whiff. And so each smell has a perceived length and duration. Both are very qualitative: smell walkers tend to agree that they smell the same thing, but not its intensity or duration. This disagreement may be determined by physical location, micro air currents, or an individual's subjective access to the smell, adaptation and habituation. So we all negotiate with others, and how long a smell lasts as a perceived part of the landscape is contested. Smellscapes (an overall composite of individual smells perceived at one time in one place) are defined by Porteous (1985) as being time-based, in that they yield experiences that are discontinuous, fragmentary and episodic.

Smell is open to conditions that affect its own temporality. But when smell is a subject, its temporalities are bound to human perception and recollection, to the messiness and vagueness of association and memory. As an object (a super-light molecular structure), smell is vulnerable to air movements and subsequent volatilisation. As a subject, the complexity and messiness increase according to genetic predisposition, prior experience and hedonic ratings.

## Sticky memories and teaching smells

*Sybille:*     Kate:                   **Chris:**

*And what about memory? Because when a participant creates a tour, the temporal and spatial story surely must be anchored to memories of smells and associated meanings? So is that another temporal dimension you're working with?*

Yes. But I wouldn't call that a history, it's more like a meshwork of multiple moments of time. The past is living on in the present, but not the whole past. They're bringing particular memories through to the fore when they are actually noting smells down. So the temporal dimension expands beyond the linearity of the 'inverse time arrow', which becomes something that is very messy and where multiple moments of memory may be simultaneously accessed and combined to form an overall impression of, for example, 'seafood'. The smell walker's association at this particular moment might have been 'rare, family dinners out to Chinese seafood restaurant. Where my dad used to take us on his rest days.' This collection of memories evoked by smell is not so much a history, more a synoptic synthesis of a series of moments united by an odour. Like chewing gum …

When I'm researching for a project collecting smell data, I'm collecting a description (the character of a smell, what it smells like), I'm collecting intensity (the perceived strength of the odour), and I'm collecting hedonic tone (the degree of pleasant or unpleasantness), expectation (surprise, or lack of, in detecting smell in a place and time), and association scale (particular, subjective connected thoughts). Recently I've added in the duration of the smell to incorporate a temporal dimension to the maps.

The last category, 'how does it make you feel? Any immediate memories or associations?', is a way of encouraging people to actually think about

**Yes – can the history of almost anyone's life be mapped out in your practice? Because surely how they have learnt to smell influences how they actually smell?**

the smellscape they are witnessing as they pass through. This mindfulness is intended to help people reflect on smells as they walk around, so it gives people a more meaningful connection with the data-collection exercise.

So in a way, smell mapping has two distinct elements to it. There is a performative mapping of a neighbourhood as a smell walk, during which data is collected, and then there's a mapping of the smell data or information about smells themselves, presented as a piece of art or design work.

*When you see people noting things down in smell notes, are there a lot of people who talk about their memories in what they write down? Or do they record feelings? Or are feelings often attached to memories of smells?*

Feelings are frequently attached to memories triggered by smells. A lot of the associations, emotions and memories that people allude to relate to childhood. Douglas Porteous (1985) hypothesises that environmental smells become less noticeable as people get older, and in a small survey of eight autobiographical works discovers most smell associations occur between the ages of eight and fourteen years.

*Is there a biological reason for that? Perhaps you smell better when you are young or you're more open to smell as you are learning about your environment through smell? Less conditioned?*

Theory suggests that the richest period of odour sensation is at the start of a child's autonomous exploration of the world, which ends with the onset of puberty. Childhood is a time when the most primitive senses are most open, before we learn social norms and behaviours. And there is a thought that kissing emanated from sniffing, and so during puberty we may start to use our olfactory sense for other purposes. You are less conditioned; you're actually doing a lot more finding out about the world at that point. And then there's another theory that says that, as you reach puberty and you're starting to smell for different reasons, it's less about leaning about your environment –

*Interesting. So smell becomes sexualised.*

And then it's more about actually smelling for other people, rather than smelling for environmental learning.

*Yes, but I am sure that certain cultures have strong informal smelling traditions which are passed on, even if only to smell whether food has gone off, for example. Mary Douglas' (1966) work on purity certainly suggest that mapping smells and identifying anomalies conveys a sense of cultural order.*

**That's very interesting because presumably smell is marginalised in formal educative practice in schools? There are very few formal curricular links in terms of the smellscapes; so I guess we simply pick things up through cultural practice rather than through formal systems.**

Yes, there are even cultures where smell is deployed to tell time. Majid and Burenhult (2014) describe how the Jahai language and culture in the Malay Peninsula is particularly rich in olfactory symbols so smell knowledge is passed on from generation to generation through culture, not pedagogy. There are cultures where smell is a means of ordering the world; the Andaman Islanders in the Indian Ocean associate smell with two very separate environments; the salty ocean, and the potent floral scents of the jungle. Their village spaces are also delineated by shifting olfactory zones based on animal enclosures and planted areas, which change according to wind and temperature that fluctuate according to daily and monthly rhythms. So here smell knowledge is passed on informally. You are right there is almost no pedagogy for smell apart from training for the wine and the perfume industries, and a growing interest from beer-makers and aromatherapy. The only times that you see smell being taught is for specialist subjects.

Smell is very rarely used within the blind community as a navigational tool to help knowing about the world. But the Fife Sensory Impairment Centre uses smells to 'name' rooms and the experiences they contain for participants.

*And tactility as well. So the auditory and the haptic can be pedagogic methods for knowing the world when the visual 'fails', but apparently not smell.*

Yes, haptics – the tactile sense and echolocation are both importantly taught within visually impaired units, but smell remains an untapped sense for the visually impaired community. In fact about ten per cent or fifteen per cent of our knowledge comes from the smell.

*You've done smell mapping walks in many different countries with many different cultural settings, like Singapore, Marseilles, Amsterdam, Pamplona, New York or Ellesmere Port. I know this is very tentative, as your research is qualitative, but do you see different patterns? Does it differ how people attach feelings to*

**Does that apply for visually impaired kids as well, people who haven't got vision? Because vision obviously is so dominant. So if you go to a school for the blind, does that mean there is formal smellscape training as a mobility aid, or …?**

**That's interesting because I know they do work around sound shadows and –**

**So smell is seen as a very individual sense rather than a social sense?**

Yes. Smells are generally regarded as highly individual, especially in Western contexts. We don't know what each other smells, but we also don't have a shared vocabulary for communicating smell knowledge, unlike in some other cultures where smell is a more intrinsic part of everyday life.

smells? Or is there a sort of continuity in how people associate certain things with certain smells?

To date I have not really analysed all the associations that people have with specific smells, but I am certain that there are both cultural differences and similarities. Smells of food are prevalent in urban smell walks, and while the names of the smells vary according to local traditional cooking, the associations are frequently family or socially based. The smell of home is another recurring association, but descriptions are wildly different; in Newport, Rhode Island, USA the smell of home is the scent of the ocean (see Figure 3.4) whereas in Ellesmere Port, UK it is the 'industrial odour' (occasionally likened to 'something burning in the oven').

*Yes, and different cultural spaces emanate different smells, which makes for cultural difference. So smell transcends any nature/culture binary. The canals in Amsterdam for example are constructed, but also evoke a natural sense of the place, and the pungent smell of Durian fruit in Singapore is so ubiquitous because it is an important part of family rituals. Chinatown smells quite different from Little India elsewhere in the city.*

I use smell associations to aid decision-making when deciding which smells to include. So whenever I create an artwork or do smell mapping, I'm looking for a representative range of smells from a sample of background smells, whether these are episodic, localised or what might be called 'curiosities'. Curiosities are the beguiling elements in a smell map, the lures and hooks to draw the viewer into contemplating smell experience in the city. Who could resist looking at a map that indicated 'dinosaur', 'deep, dark secrets' and 'durian' as its curiosity scents? For me, the curiosities are what comes from the associations.

**Which are local, presumably?**

Which are local, but also individual, emotionally influenced and associative. So that is where we get the unexpected scents and the ones that actually make the smellscape so

Mapping the quixotic volatility of smellscapes

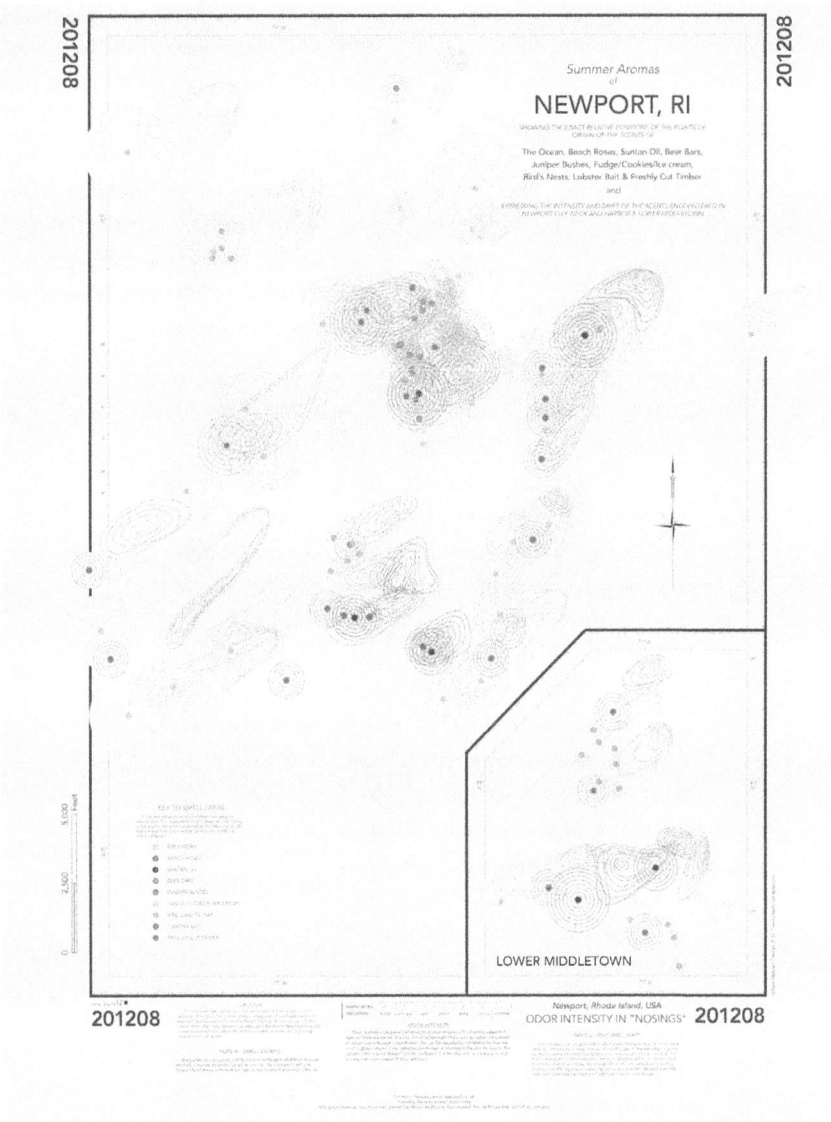

**Figure 3.4** Smellmap Newport, Rhode Island © 2012 Kate McLean. This figure has not been made available under a CC licence. Permission to reproduce it must be sought from the copyright holder.

different from odour-monitoring practices and quantifiable measures. This challenges accepted norms – which I see as a critical part of my art practice.

So it would be fair to say most of your work is focusing on current practice rather than past?

Yes.

Your concern with memories is in so far as they alter what people actually do, rather than just being concerned with recollections and memories?

Precisely. My work is wholly based in present perceptions, the evolving nature of urban smellscapes. One very early piece of work, 'Smellmap: Paris' (see McLean, 2010), was an exploration to see if associative memories might be prompted by specific smells, and whether a transferable, universal urban smell existed. What resulted from this experiment was wholly unexpected and a testament to the uniquely personal associations with specific smells. It asked the question whether people smelling something conjure in their own memories and imaginations a place or an emotion. I created a set of fourteen scents particular to smells associated with Paris. What I thought was that the smell of cigarette butts might entice people to write down the name of another city, that the smell would take them to memories of Brazil or Rio or Milan or Prague. But this didn't happen. Instead people wrote short sentences like, 'it's a railway platform late at night'. So when we talk about place in terms of smell we are saying something intimate and personal, coming from subjective memories.

And it's about an experience rather than a place. It's about an event, how that story relates meaning in that context. Not space or place.

But what about the other direction? What about going forward? We talked about memories a bit. What about futurity? What role might smell play in your work in terms of bringing possible futures into being, as against recollecting pasts?

In terms of bringing the future into being a potential role for smell is to think of it as a communication tool in its own right, more than simply a navigational aid. As human beings we're very extraordinarily good at processing vast quantities of information. Bushdid *et al.* (2014) suggest we can discern up to one trillion different smells; so we are already dealing with large amounts of data and data processing.

Perhaps we could actually use the emotional aspects of smell to communicate data and to relate the spatio-temporal qualities of being? Using smell as a communication device in its own format, I think, is really interesting.

Yes, or you could walk through a city and choose a route based on smell, or a lack of smell. Or you might send a friend a smell walk, so as to connect your world with theirs. The app that I'm developing at the moment will enable people to set their own smell walks. The idea is that users would be able to indicate moods via a pointer so as to suggest experiences.

**So you might know the city in the future in different ways by presenting somebody with a smell map; you almost call an alternative city into being in the future because you can choose to avoid the fish market or go to the fish market?**

## Smellscaper

*Sybille:*

Kate:

The Smellscaper app is based on OpenStreetMap data and I want it to be open source for both development and use, so that anybody can then take what we're collecting. There is a large database behind a digital version of a smell note which is capturing the smellscape (see Figure 3.5). It will also geolocate every single smell point, down to

**Chris:**

**So picking up on that, tell us a bit more about the Smellscaper app and how that technology alters temporality.**

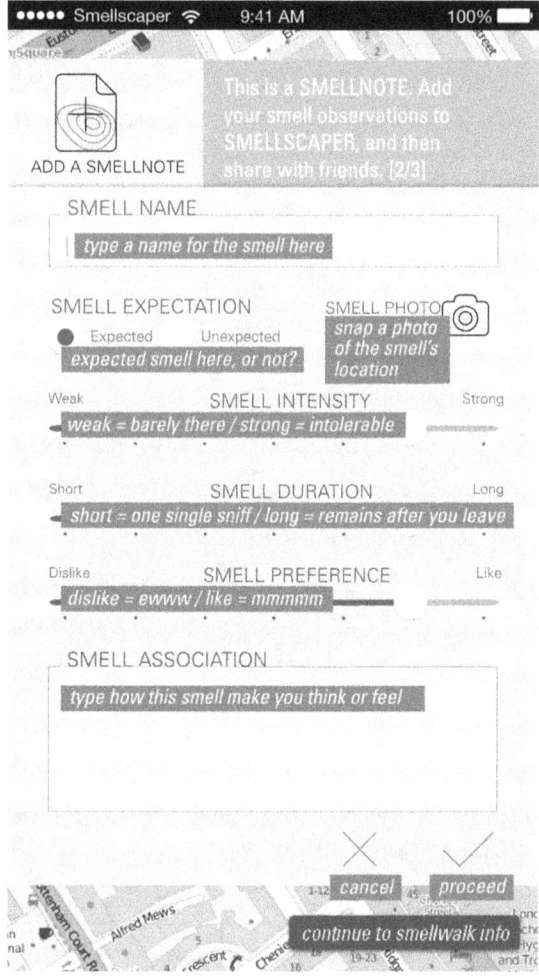

**Figure 3.5** Smellscaper App Smell Notes interface (courtesy of Kate McLean). This figure has not been made available under a CC licence. Permission to reproduce it must be sought from the copyright holder.

about a building's width. And simultaneously the Application Program Interface (API) lets us capture weather data at the point of smell note completion, particularly wind and temperature.

I'd like to create simultaneous, dynamic mappings of several cities. So we can then look in different ways at multiple smellscapes from a wide variety

Mapping the quixotic volatility of smellscapes

of global locations and see what emerges. The app is currently being developed as web-based – ideally it would be available for Android and iOS platforms. Users could compare smellscapes in different parts of the world at the same time.

*You retrieve all the data then? And then although it is free and open, you can use it for your own practice, to access a rich collection of smell data for your own mapping practice from different temporal landscapes.*

Then you crowd-source the data; it pulls the data back and then you can chart multiple temporalities, and how they may relate to spatial stories in particular places?

*Temporal and spatial stories again. So what is it called again? Smellscaper?*

Smellscaper.

## Rhythms and durations

*Sybille:*

Kate:
Absolutely! I'm just starting to think about the rhythms of smell perceptions and how they happen. In the piece of work that I did in Pamplona, it was very clear that each of the walks had a rhythm of its own which revealed aspects of the rhythms of everyday life. People participating said, 'we are incredibly concerned with food here', and their collective understanding of their own pace and … there was one smell note from there which was 'place: bakery, smell: nothing, comment: absurd'. And within that there is the expectation of 'here is my smellscape', knowing what to expect within this and then suddenly

Chris:
There's the temporality, the sequential temporality of the smell walk, but then there's the individual temporality of a particular perception of a smell duration. But presumably there are also rhythms, which is not something we've talked about yet.

So in the light of that, what is producing the temporality? Is it the smell or is it the everydayness of life? I'm thinking that in some ways it's the rhythm of life nothing happens. This changes the rhythm of the walk.

**that is making the smell experience, rather than the rhythm of smell. So it's kind of a question about how smell relates to life.**

*Smell is life. But we seem to forget that life is perceived through smell.*

It is life producing the temporality, rather than the smell ... but the perception of the rhythm can arise through olfactory awareness. Smell perceptions congregate on street corners which index a confluence of people and movement. Street corners are where you get a confluence of people coming in; there's much more activity. As smells index life, urban smells index city rhythms – there is much more to discuss here. It's definitely life.

**So it's about encounter. So in that way smell becomes a social sense. It doesn't become an individual sense in that case?**

*It's both. I mean, as Kate shows in her work, smell can be highly individual and attached to very personal associations, which may be totally different from someone else's. And also we sometimes lack the words to share it in a social fashion.*

*I just wanted to ask something about how long smells last. When you talk about smell duration, how does that relate to ephemerality of smell? I mean does the ephemerality of a smell directly correlate with its duration. Is it the same?*

Smell duration is the perceived duration of an encounter with a particular smell, it is human-sensed. Ephemerality, on the other hand, is of a different order. It is an attribute of the physical properties of the smell and the environment in which it exists. If participants move the duration of a smell may be limited, yet the smell itself may not be ephemeral.

**Perhaps we can see this in some of your animated work? What about your dots diffusing across Amsterdam?**

I think the Pamplona work explains this better.

To explore possible visualisations of ephemerality, I initially printed onto a translucent substrate. But time-based media and animations somehow seemed more like what I wanted to capture. The motion graphic of 'Smellmap: Amsterdam' depicted a collective smellscape coming into being, existing and shifting for a moment in time, and then incrementally diffusing as it volatilises and disappears. In 'Smellmap: Pamplona', I combined this diffusion with the contestable nature of smell. My interest there was to explore what happens when you've got a smellscape with individually varying smell temporalities. This piece explored the dynamics of the smellscape, reframing time to reflect a sequence of individually perceived smells. What happens when somebody else comes in and moves somebody else's smell out of the way? I'm looking at specific sections of the city: a smell comes in, drifts away and volatilises, and then a separately perceived replacement arrives. Another one takes its place. That seems to be closer to what actually happens – at once individual, moving, ambiguous and changing.

*So that immediately relates to ephemerality and duration as well? In this case the participants are not mobile, so do duration and ephemerality directly correlate? I mean, when participants are on the move, ephemerality is also influenced by the mobility of the walkers. They, so to speak, leave a smell behind.*

Yes and no. Some participants were static; others were moving. These smells are instances of perception. That really relates to ephemerality. So this is what that looks like.

'Smellmap: Pamplona' starts with the geographic features of the landscape which appear to give a context to the smellscape: first the river, then the transient and negotiated green spaces in the city, and finally as the park areas disappear, the city's urban infrastructure of roadways is mapped out. Gently these fade into the background and the scents start to emerge. The first smells in each quadrant appear, but each one has an individual life, rather than the collective life of the smellscape as portrayed in 'Smellmap: Amsterdam'.

I've got data in a spreadsheet from about fifty people.

**So tell us, how does that feed back from the data which is generating this visual? Have you got data in a spreadsheet, from individuals who've experienced that?**

I've got about fifty people that have smell walked their city and contributed about 420 individual smell notes. They each located their notes on a gridded map of the city. This gave them coordinates, and any smells that were located became the data for the map. For a month prior to me arriving, two maps had been on display in public areas and individuals had started the process of contributing written individual smell notes. These smell notes contributed information about location, name, intensity, hedonic scale, duration, association, link and whether or not this particular instance of the smell was expected. These smell notes were translated into Spanish and pinned to a gridded map of the city, then subsequently translated back into English for me to work from. In the end I selected 121 individual smell instances and animated each one separately based on the participant's perceptions and local wind data. We then carried out five smell walks over four days and collected ridiculous amounts of data, and then I took each of those, animated its life form and combined each of these together to create the map. The sequencing is based on the order of perception within individual grid squares in the map, so these are in sequential order and as smells come in, one replaces the other. So time is asynchronously warped and extended in different parts of the city. Temporality shifts. It's changed time in the process because I'm

**Okay. And they have walked across the city?**

Mapping the quixotic volatility of smellscapes

not concerned with absolute chronologies or real-time. It is more of an olfactory sequencing in space and smell time. In other words, do certain smells come up in certain places? These were categorised and classified in terms of a smellscape categorisation. So, there's a tree for example.

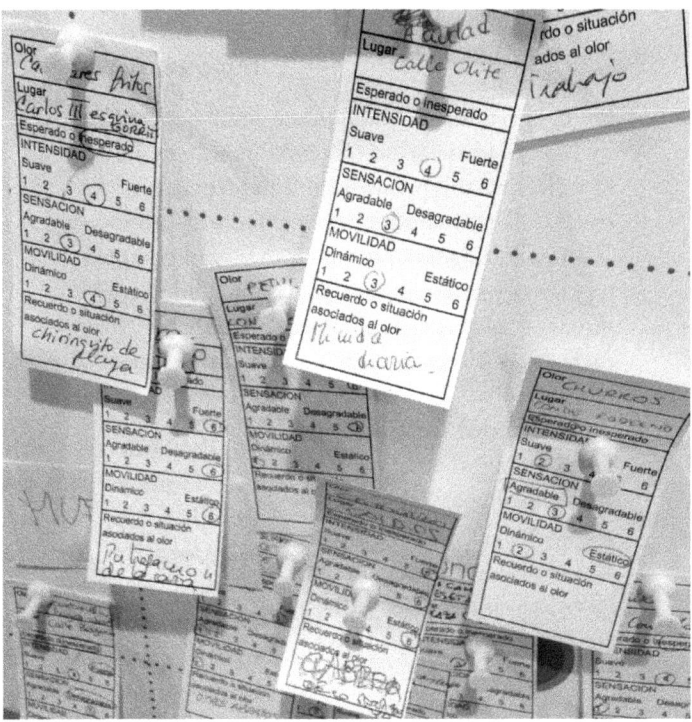

**Figure 3.6** Spanish language smell notes prior to classification (courtesy of Kate McLean). This figure has not been made available under a CC licence. Permission to reproduce it must be sought from the copyright holder.

To allocate smell colour I categorised the smells according to Victoria Henshaw's (2013) urban smell categorisation. So there's traffic emissions, industry, food, tobacco smoke, cleaning, synthetic waste, people, animals, nature, buildings and non-food.

And those were the ones that were identified by

*So this maps a complex and shifting meshwork of particular smells.*

Henshaw in her book and PhD research on smellscapes. And then as a result of analysing the data, I realised my practice elicits two further categories of urban smells. In order to be able to map it in the way that I wanted to, I added 'emotion' and 'complex' as categories. So the 'smell of an exam' is not actually something that fits into an urban smell category, but it's part of the overall urban smell experience.

*So these categories are all based on objects? Or things? A factory, for example?*

**Whereas olfactory perception may derive from feeling or experience.**

That is what the smell notes indicated, and so in reflecting the practices and perceptions of the smell walk participants, I decided to add 'emotion' as a smell category. And the category of complexity occurs when many smells combine into a single moment of perception, indistinguishable from each other at a single sniff. This is another aspect of temporalities being evoked through smell. Lots of the smells that we experience when we're walking or when we're being asked to think about smells are complex: you realise, one sniff at a time gives you something but there are contextual features to it too. So a single smell experience may mean that people come up with complexity as well as any individual smell note when they're writing them down.

It depends on the person perceiving the smell, the physical characteristics of the smell and the environment as well.

**So I've got two follow-up questions to that then: the first is a kind of empirical temporality question; the second is a methodological issue about the temporality question. The first one: when we started out, you were talking about recording the ephemerality of different things and moments. How does that vary between these things? Will there be certain smells that have a longevity, and certain smells which diffuse immediately? Does that depend on the smell or does it depend on the context?**

*And the weather as well.*

*Would you say that the world smells differently today from the way it used to? Do you have any idea about what changed about the smellscape in urban environments?*

The weather is a contributory factor in my smellscape mappings. And I'm taking individual pieces of data as being what I am basing the map on.

There is work around that. So if I go back to the urban smellscape pyramid again, Porteous (1985) in his 'smellscape' paper, he describes 'background' or 'ambient' smells that give a general context to a place, a smellscape. And then there are 'episodic' smells that are specific to time and specific to activity. So, for example, a market will generate episodic smells from the stalls. A restaurant will produce episodic smells of cooking. The back alleyways behind a restaurant with extractor fans will experience episodic smells that are different from the ones that appear inside, which are again different from those that escape through the front door. The rhythms of shops putting flowers out, people drinking in cafes, smoking, will each generate episodic smells which are usually quite localised, and you can imagine might really characterise a city when they are sensed.

Classen, Howes and Synnott (1994) describes

But I'm thinking, we're by the sea at the moment. There is a pervasive, vague, 'seasidey' smell which is here always and never really goes ... it changes, the intensity changes, but it's always kind of here. Whereas there may be a very specific, I don't know, the smell of a meal. The question was in the sense that, while it may be individual, there are certain smells which diffuse and certain smells which linger, and whether there is work around that.

the smells of antiquity, explaining the division of Roman cities into trade-related areas that each had their own aromas, as well as the overall scent which was far more perfumed than contemporary cities, possibly to cover the fouler aspects of life. In some ways this hasn't changed. Our recent paper about people tweeting smellscapes reveals that different parts of London, for example, are more likely to be perceived as having associations with one smell as against another (see Quercia *et al.*, 2015 and Figure 3.7). However, Classen, Howes and Synnott (1994) suggests that the gulf between our deodorised lives and that of ancient history is deep and wide. If you go back to Roman times, then cities were very smelly places before we had waste treatment.

**Figure 3.7** Contrasting smellscapes of the West End of London, mashup showing emissions, nature, food and animal smells against an OSM backdrop, as recorded in social media (courtesy of Schifanella Rossano, http://goodcitylife.org/smellymaps/). This figure has not been made available under a CC licence. Permission to reproduce it must be sought from the copyright holder.

*Some Belgian cities still had open sewerage for a long time.*

**And the French outdoor *pissoirs* survive to this day …**

Yes, so cities were very smelly places. However hard we try to eliminate smell it remains a feature of urban living. And there still are cities that reek, especially as temperatures rise; New York can emit some pungent odours comparable to Singapore; Paris on a stifling day can match Valparaíso. This depends very much on how we deal with trash collection, waste and effluent. But when industry was more located inside cities, then there were very episodic smells coming from those industries. In the UK prevailing winds made a big difference to where industry was sited in relation to residential areas; London's prevailing westerlies ensured the West End remained far distant from the polluted air of the East End. Gentrification happened in part because the western side of cities were generally regarded as nicer places to live, because there you didn't have the smells from industry drifting from the eastern side, which is where the poorer communities lived. And what's happened now is that industrial smells generated by things like breweries are now moving out of the cities.

*Except in Amsterdam!* Our urban smellscapes are changing, industries move out, gentrification with its attendant aromas moves in. Yeah. And that's changing urban smellscapes, and this is partly why I am recording the data, to give some sort of tangible form to recording smells, in the same way that we can record the visual image of the cities.

**So, we will come back to the methodological question in a minute, but in that case, does the smellscape flow almost from the mode of production? As capital alters the world, does that in turn alter the smellscape?**

Yes. And there's some work by Alex Rhys-Taylor of Goldsmiths College, London, about this ... he walks from Shoreditch into central London and locates political change through altered smellscapes (see Rhys-Taylor, 2015). In that walk he says you can smell gentrification. Street food has become a large part of the urban smellscape in the UK as the evenings witness outdoor bars sharing their clientele with gourmet food shacks. And there are acceptable street foods and unacceptable street foods. Rhys-Taylor (2015) says that as you walk through Shoreditch, you're in a diverse and changing smellscape, full of the shops that are associated with gentrification, and in the daytime the whole of the hipster movement generates its own particular smellscape.

*Roasting coffee beans and the smell of sourdough.*

Yeah, the sensory landscape becomes a large part of that hipster world experience. As you head westwards towards Central London, the shops that can afford those rents are the ones that have smells that they've identified and generated for themselves. So individual shops give way to those that can support the higher rents. Abercrombie & Fitch, Lush, McDonald's, Burger King, Subway, all have their own unique fragrances designed for marketing purposes. The smellscape has become commodified. So you hit a very different commodification of those smells and their consumption.

**But in one sense that is kind of the built form generating the smell, but you've also got the individual lifestyle carrying a smell with it. So the hipster will have beard oil as a particular scent …**

*Roast coffee beans and the smell of sourdough.*

Coffee beans, artisan coffee beans. Baking bread. Beard oil. Skin moisturiser. Produce teasingly directing us to eat and drink combines hedonistically with olfactory bodily adornments. And the importance of putting that out to the street as well. I know supermarkets will push the smell of baking out onto the street, although they're not baking in the supermarket.

**Mostly we've been talking about the smell out there rather than your apprehension of the smell. And we've talked about the past and we've talked about the present and the future, talked about the**

rhythms, talked about the ephemerality. What we've not talked about is the processual temporal nature of your research and the extent that you're producing answers instead of discovering them. You are carrying out a piece of research that has a fixed point at the end and there's a trajectory to that; it's time's arrow and you're going to be fixing something that is constantly mutable and changeable into an apparently stable map. Could you tell us a bit about how time impacts on the process of researching smell, and how you choose what you choose to do? When and why that might matter to the kind of results that you get. Sorry, that's a difficult question!

The process that I go through with this is very intuitive, and in particular how I actually move from one type of mapping to another. So each smell mapping that I do then makes me think about different elements that I want to consider. So I realised with Amsterdam, for example, that I wasn't happy with the idea of the smellscape being an impressionistic capturing of a moment in time.

It was frozen. It was there. It was a moment and then it dis- **Frozen.** appeared. Similar to how the French Impressionist Claude Monet captured single moments of light on Notre Dame Cathedral. I wanted to do something more with that smellscape than say, 'here it was for that particular single moment'.

Exactly. And so I didn't feel that was actually being true to it. There is a huge degree of interpretation on my part; I select the final content to put on the map. There are many omissions ... and ephemerality and temporality became really important to me. Smells disappear in an instant and then occasionally re-emerge. One of the really important pieces of research from Amsterdam was listening to a recording of somebody doing a smell walk where they said, 'Ooh, it's there. Oh, I've just lost it. Oh, it's come back again.' And that to me suddenly became the focal point to communicate: ever since, I've wanted to create maps that show how smells disappear. This ephemerality is a particular quality of smell which makes it so difficult to map and needs to be incorporated as integral to the mapping. So with Amsterdam, I said right, okay, let's see if I can indicate that temporality. And in the end all the scents evaporate and dissipate into nothing.

**Because it's not a single moment. Effectively it's multiple moments that are pretending to be a single moment.**

## Methods and methodologies

*Sybille:*

*We were talking about the process of you designing and doing a smell walk, and we ended with the moment when you describe how you prepare participants to smell in a particular or different way.*

Kate:

**Chris:**

I do try to prepare them to smell in a different way from how they would normally. My methodology is called 'walking nose-first'. It's about experiencing the world nose-first, changing from the ocular-centric to the olfactory-centric. Registering the everyday smellscape is a very important part of the practice, so doing the smell training is as much about how and where to sniff, as about the types of encounters that you might want to have as you're doing it. Initially when I ask people to sniff inside bins, there's a level of reticence but this tends to evaporate once I frame it as art practice and people are quite willing to perform activities outside their normal comfort zones.

Very often smell walkers are surprised at the sheer volume of smells they notice and how many of those are more pleasant than they thought. Our association with smell in contemporary society is more generally with malodour than with odour. We complain, and odour monitoring is called in to determine if odour control is required. We tend to like urban smells in restorative environments such as parks and gardens. But the everydayness of smells is only very rarely acknowledged, so one of my aims is to try to get people to pick out the smaller smells of everyday life.

**Can I pick up on that and ask you a methodological question? Are you creating the smellscape which you want to create, rather than apprehending the smellscape which is out there? So by instructing people and carrying out some kind of action research, are you altering the world?**

I don't know ... when people are actually recording their smellscape, I don't involve myself at all from the moment they start to the moment they finish.

I'm directing their perception and thinking towards under-explored ways of knowing – so yes, to a degree I am changing the world by guiding them to physical, perceptual and mental spaces that they wouldn't normally think about. But at the same time I have absolutely no control about whether they decide to record those smells or not.

**But you're still setting it up ....**

*And is that a methodological choice as well? I mean you are not just an observer: methodologically speaking, you are sending them in specific smell directions, taking them out of their comfort zones or routines.*

Absolutely.

*How long does a smell walk take on average?*

About forty-five minutes; concentration and focus diminish after that.

*Do you notice people getting tired more quickly, perhaps because they're not used to using those sensory abilities?*

Yes, exactly. It is that. You get physically worn out by sniffing unless you pause to break. Humans have a limited capacity to smell deeply for any length of time; a process of adaptation ensures that we take olfactory breaks and stop perceiving smells. As we stop being aware of every new smell, so interest drops. What usually happens is at the start of a walk, you'll have a lot of smells recorded, and then it gradually decreases. Recently I have devised a new approach which segments the walk into stages, and I've found it beneficial to change the rhythm of the walk to ensure a more even distribution of smell perceptions.

**Okay, so, picking up on that, your approach and methodology is very often about mobilities and also that mobility of temporality, in the sense that they're going on walks; presumably you could do it through auto-mobility or through stasis, or being in one place?**

Sitting still and waiting for smells to come to you.

**Encountering the smell and searching for the smell. Being fixed will actually apprehend a different kind of smell.**

Absolutely.

**Which would be interesting if you were able to choose the right – I'm not sure 'right' is the right word, but you know what I mean – places that would work.**

I've experimented briefly with fixed smellscape mapping in Newport (Rhode Island), where I sat on an Adirondack chair for a couple of hours and waited for the smells to come to me. I now sometimes ask smell walkers to perform this 'smell catching', where smells on the breeze are apprehended and noted.

*Can we get back to the process? Because we've ended up with you preparing, and then you send them off and then you leave them, and it's their process. And at the end of the day you all gather together again, right?*

They walk and catch smells, stopping after fifteen minutes to compare notes, perceptions and observations. Then I ask them to use other senses to seek out potentially interesting smells – 'smell hunting' – for a further fifteen minutes, and we stop again to consider anomalies. The final stage is 'free smelling' before a plenary sharing session, in which I collect their smell notes and they depart having undergone a new experience in a place that they think they know.

*But you don't give them hints; you're more of an observer ... participant observation?*

Yeah. I walk with them but let them be. If they have any questions I answer them. At the end we usually talk about the experience collectively, about what did they get from it, and what did they notice. For example, in Pamplona, people were saying, 'we've just realised how important food is as a part of our culture in the city'. And I'm just starting to consider this as an important component of the walk. To date I've been terrible about drawing them together at the end; I've just let people go, and I've just started to have enough confidence in my practice to pull people together and ask 'what did you get from the experience?' Respondents have noted that they see the city in a completely new light; it's slowed them down; they've seen things that they hadn't seen, as a result of smelling or searching for smells. So they've noticed aspects of the city that weren't previously in their world view.

*And sometimes you let them make their own maps and sometimes you don't?*

Sometimes I let them make their own maps. More often I collect the data for my own mappings; however, occasionally I'll be asked to run smell walks and mappings as events, in which case the mapping becomes a part of the participant experience. Cars, dogs and the smell of rain to come.

**It's interesting, the extent to which our technologies have altered the temporality. The car has been part of sanitising the smellscape in that sense because it removes you from the smells.**

The car has had a triple effect: first, as you say, as a distancing mechanism, but at the same time as a commonly noted urban pollutant to walkers, and the smell from car exhaust also masks the other smells in the city smellscape.

So you're not actually letting smells in through the window. The ideal car for a smell mapping is actually a Cabriolet.

**And the way in which increasingly cars deliver a sanitised safe private space.**

**My question was more about how temporality gets changed by technology and the role of speed. If people go on a smell run, is there a different perception of a smellscape from people going on a smell walk. And if they're all sitting still, is that a different perception from moving?**

Research is yet to be conducted. But I think so. From personal experience, when I'm running I'm moving through a landscape at a pace, and it's the impact that individual smells have that forewarns me about what is coming up, or which tells me about the place that I am in. So as I'm running smell can act as a 'pre-vision service'. I will know that there is a farm coming up.

*That is interesting because we have one question about you predicting smells, futurity? So you can predict smells by seeing things?*

You can predict smells by seeing things. You can also use smell to predict what you are going to be seeing.

*Yes, my dog seeing me on the screen while I was Skyping my daughter and then going to the window, trying to capture my smell, as if I was on my way home.*

**It's like your anecdote yesterday, Sybille, about the mismatch between your dog's visual perception and olfactory perception.**

The other day I was running up from the village. There was the smell of manure and I knew that very soon, because of the way that the wind was blowing, I was going to be coming across either a farm or a horse. And in my own mind I was trying to figure out, do I think it's cows, or do I think it's a horse that is actually creating this smell? And I turned right and there's a horse in a field. The pre-visibility that smell affords is really interesting.

*Also, if you see the weather forecast you can predict smells and you probably do that to a certain extent subconsciously as well. When there is rain forecast after a very hot day, you can already imagine the smell of rain on a dry pavement and the sort of dusty smell that gives off.*

It's called petrichor, which is an incredible smell on its own; people call it 'the smell of rain'. But it's what happens when rain hits dry dust, with bacteria, and that then releases the smell. Petrichor pre-empts the rain, as it is released with the first moisture droplets to hit the ground. The finer the moisture, the more intense the scent.

**Forecasting things changing, isn't it? The weather forecast was one of the first, but then you have pollen forecasts, you have pollution forecasts. You kind of see the way in which the functional comes into the artistic. You may end**

This is a direction I'd love my work to take. I'd like it to feature on screens in a similar capacity to a weather forecast. A motion graphic visual of 'tomorrow's predicted smellscape for your neighbourhood', with an iconography in the form of criss-crossing isolines indicating where local smells may be found, predictions of possible smellscapes that you can only ever know by being present in them.

up with an aesthetic smell forecast. 'Today is going to be a very good smell day.'

It's interesting, the way in which forecasting has kind of changed. You used to have the isoline chart and the synoptic weather chart and now often you have a much more functional indication of, say 70 per cent chance of sunshine. And how you target the visual translation of a smell forecast to make it readable for someone is really interesting. That is an aspect that we haven't really talked about, how you make that translation between modalities.

Maybe we should talk a bit about your choice of design, visual design variables, such as colours and how they work in your practice.

Colour ... changes depending on the urban smellscape that I am mapping. So I pick the colours that I'm going to be using as a visual reference from the fabric of the city itself. I eyedrop the colour of the sky and the hue of the trees, developing a palette of the natural environment. I consider paint colours used for external walls, window frames and doorways; I sample the colour of cars, clothes and graffiti, pavements and walkways, signage and railings.

The palette is not based on a single hue or intensity; smell colour and smell intensity are not related; the only imperative is that they work as an ensemble. There was a point in time when I picked the colour based on reference from the smell source, and to an extent this still happens. My only exception to this was Newport, where I struggled before finding a colour swatch that worked for the city.

Yes, yes, absolutely. And that all depends on whether sewage features in the urban smellscape.

**The palette is based on hue or intensity or ... or what? It varies?**

**So in one city you might find that sewage might be brown and in another city it might be blue.**

**Which is interesting because it kind of speaks to the differences between art and science in terms of how you depict time. In one domain, knowledge is profoundly different from how it is depicted in another. Would you say that you're more of an artist than a scientist? Where do you place yourself?**

An interesting question because when I talk about my practice it comes across as scientific and analytical – like a data visualisation – yet methodologically it draws from the social sciences, whereas in the process of creation it is both design-oriented and artistic. I find smell to be a discipline where art and science collide and this is a constant tension and reference in my work.

## Multiple maps, multiple moments, multiple dimensions

*Sybille:*

**Kate:**

I am convinced that there are multiple answers. It has always been an objective of my practice to put the process in the public domain and enable a range of approaches and viewpoints along with the mappings of urban smellscapes. Comparing different mappings is work for the future. Initially, I considered keeping my practice private and secret, but some smell practices are already considered secretive and specialist (for example in the wine and perfume industries). Since I deliberately engage with a general public and require no specialist knowledge, I feel it is important for everything to be as transparent as possible, and that includes techniques and processes. All I ask is that it is reciprocal. I believe that with increased interpretation of the urban smellscape, patterns will emerge, commonalities and differences will appear and we will start to comprehend our very human relationship with urban olfactory landscapes.

Yeah, so as far as I'm concerned, the practice is open source. If somebody else wants to go out and do it as well, I would love to see the results.

**Chris:**

So hypothetically, in the brave new smell future, you've got multiple imitators, with different people making different claims on the smellscape, with an alternative competitor saying, 'hey, my maps are better than yours', doing it in a different way. How do you react to that, with different temporalities called into play with different people? What I'm saying is, is there one answer to mapping the ephemeralities of a smell, or are there multiple answers?

And you don't know how your practice is going to change in a year's time because you don't know what the smellers will enjoy and how that will alter your practice?

Traditional maps are rendered in two dimensions; digital technologies afford us the options of mapping in four dimensions, but I am not convinced that even this is sufficient to map human-centred urban smellscapes, when Magnasco, Keller and Vosshall (2015) suggest that humans can discriminate up to one trillion odours and that it is possible that olfaction operates in 400-dimensional space. There is plenty of opportunity to conceptualise mappings that extend beyond the existing spatio-temporal paradigm.

*Thanks for taking the time to talk us through this Kate! It's been fascinating.*

## Acknowledgements

Chapter design by Kate McLean.

Due to the limitations of the print format, some detail has been lost from the images contained in this chapter. For high-resolution colour versions, please see the Open Access edition at http://doi.org/10.9760/9781526122520.

## References

Bissell, D. and Fuller, G. (2011) *Stillness in a Mobile World*. London: Routledge.

Bushdid, C., Magnasco, M. O., Vosshall, L. B. and Keller, A. (2014) Humans can discriminate more than 1 trillion olfactory stimuli. *Science*, 343(6177): pp. 1370–1372.

Classen, C., Howes, D. and Synnott, A. (1994) *Aroma: The Cultural History of Smell*. London: Taylor & Francis.

de Certeau, M. (1984) *The Practice of Everyday Life*. London: University of California Press.

Douglas, M. (1966) *Purity and Danger: An Analysis of Concepts of Pollution and Taboo*. London: Routledge.

Henshaw, V. (2013) *Urban Smellscapes: Understanding and Designing City Smell Environments*. London: Routledge.

LaBelle, B. (2015) *Background Noise: Perspectives on Sound Art* (2nd edn). London: Bloomsbury Publishing.

Magnasco, M. O., Keller, A. and Vosshall, L. B. (2015) On the dimensionality of olfactory space'. bioRxiv. [Online] Available at: www.biorxiv.org/content/early/2015/07/06/022103 (accessed 3 August 2016).

Majid, A. and Burenhult, N. (2014) Odors are expressible in language, as long as you speak the right language. *Cognition*, 130(2): pp. 266–270.

Makin, S. (2013) Sense of smell has a genetic flavour. *New Scientist*. [Online] Available at: www.newscientist.com/article/mg21929293.100-sense-of-smell-has-a-genetic-flavour.html#.UsBtbOhL9Q0 (accessed 29 December 2013).

McLean, K. (2010) Smell map narratives of place – Paris. [Online] Available at: www.nanocrit.com/issues/issue6/smell-map-narratives-place-paris (accessed 3 August 2016).

McLean K. (2017) 'Communicating and mediating smellscapes: The design and exposition of olfactory mappings'. In: Henshaw, V., McLean, K., Medway, D., Perkins, C. and Warnaby, G. (eds) *Designing with Smell: Practices, Techniques and Challenges*. London: Routledge, pp. 67–77.

Ng-Chan, T. (2016) 'Detouring the map'. In: Driver, S., Stone, K. and Patenaude, M. (eds) *Engaging Affects/Thinking Feelings: Social, Political and Artistic Practices*. Newcastle upon Tyne: Cambridge Scholars Publishing, pp. 62–78.

Nold, C. (2009) Emotional cartographies: Technologies of the self. [Online] Available at: http://researchbank.swinburne.edu.au/vital/access/services/Download/swin:23172/SOURCE1 (accessed 3 August 2016).

Porteous, J. D. (1985) Smellscape. *Progress in Human Geography*, 9(3): pp. 356–378.

Quercia, D., Schifanella, R., Aiello, L. M. and McLean, K. (2015) 'Smelly maps: The digital life of urban smellscapes'. In: *Proceedings of the 9th International AAAI Conference on Web and Social Media (ICWSM)*. Oxford University, 26–29 May. [Online] Available at: http://researchswinger.org/publications/icwsm15_smell.pdf (accessed 3 August 2016).

Rhys-Taylor, A. (2015) Phewf, London smells! Here's the fragrant history of five areas. Now. Here. This. The *Time Out* London Blog, 26 February. [Online] Available at: http://now-here-this.timeout.com/2015/02/26/phewf-london-smells-heres-the-fragrant-history-of-five-areas/ (accessed 7 August 2016).

Rosenberg, D. (2012) *Cartographies of Time*. Princeton, New Jersey: Princeton Architectural Press.

Tuan, Yi-Fu (1977) *Space and Place: The Perspective of Experience*. Minneapolis, Minnesota: University of Minnesota Press.

# 4
## Seasons change, so do we: heterogeneous temporalities, algorithmic frames and subjective time in geomedia

*Pablo Abend*

> The frozen circumstances of space only come alive when the melody of time is played.
> (Thrift, 1977: 448)

## Introduction

As distinct from the moving image, it has been argued that the purpose of maps is to codify spatial knowledge in stagnant form by affording a firm coupling of standardised cartographic signs. Bruno Latour treats the map as an archetypal 'immutable mobile', or more precisely as an 'immutable and combinable mobile' (Latour, 1986; 1987; 2009), by which he terms a (graphic) thing/vehicle that is the outcome of scientific knowledge accumulation, the translation of this knowledge into a sign system with standardised projection techniques, and their inscription on a transportable carrier. 'Immutable mobiles' are part of a cycle of knowledge accumulation that Latour describes using the example of an expedition: the first successful expedition brings back a map of a previously unknown territory; a second expedition can build upon this knowledge to accumulate additional facts; a third expedition builds upon the second, and so forth. In order to ensure the establishment of accumulation cycles, objects have to be created that are 'mobile but also immutable, presentable, readable and combinable with one another' (Latour, 1986: 7). As a precondition of this shifting of knowledge, the immutable mobile eventually affords the exercise of power over distance in a process that renders 'facts' durable through what he terms the 'strategy of deflation' (Latour, 1986: 3). The 'strategy of deflation' signifies the production of two-dimensional inscriptions out of accumulated scientific facts by translating aspects of the three-dimensional and dynamically changing (geographic) world onto a flat and static surface. Therefore, only through fixation can knowledge be stabilised and safely brought back from distant places to central agencies Latour calls 'centers of calculation' (Latour, 1987: 215). Because knowledge accumulation demands an immutability of the sign plate, fixed maps have been mobile and thus time-critical, merely in the sense of circulation as an artefact. One consequence of this mismatch between stabilised and immutable knowledge, as against depicting the dynamics of change and history, could have been that the depiction of time has been a neglected topic in geography.

While Latour sketches the role of graphic inscriptions in the epistemological groundings of Western science, others have taken different critical positions towards the ontic status of mapping itself. For example, Doreen Massey (2005) describes how space in (Western) cartography became a static representation; cleared from uncertainties and procedural knowledge, human movement and displacement. She asks what it means to travel across space and argues that it is not an activity of simply crossing through space, but an act closer to a co-production:

> you are not just travelling *through* space or across it, you are altering it a little. Space and place emerge through active material practices. Moreover, this movement of yours is not spatial, it is also temporal. (Massey, 2005: 118)

The critique implied here claims that by dismissing any form of movement, the temporal dimension is neglected by cartography because only knowledge that can be brought into stable forms is taken into account – despite the fact that movement through space and change over time are identified as the productive driving forces behind the production of space (Turnbull, 2007). Tim Ingold calls this the logic of inversion, a central process of modern thought which turns pathways into boundaries, movements into point to point connections, and wayfaring into transport. In transport, movement is 'decoupled' in perception and motility, and becomes pure displacement (Ingold, 2009).

However, the material representations of geography have always had more or less hidden temporal aspects, even though the act of mapping has also been as concerned with time as it has with space. In addition to being mere 'snapshots in time', maps have been equipped with time stamps indicating the historical state of the world they refer to, or have been animated to depict time-critical processes such as human rhythms. But all in all, the cartographic depiction of time in pre-digital, modern times prioritised static representation of temporality, with the line being the dominant figuration of chronological information, ordering time as linear and continuously proceeding (Rosenberg and Grafton, 2010).

This has clearly changed with the contemporary hybrid media forms of the digital age, which offer new and multiple ways of time integration. Therefore, this chapter argues that one of the main distinctions between analogue maps and digital geomedia[1] (Döring and Thielmann, 2009; Felgenhauer and Quade, 2012) can be found in the way visualisations organise the temporal dimension. In order to show this, I do not reject the idea of representation and inscription, but rather look at the transformations of these concepts in and through digital geomedia. Accordingly, the following pages highlight several stages of time integration in digital maps and globes to show a transition that could be part of a larger paradigm shift in cartography: from static representations to the dynamic presentation of space, at the intersection of mobility, visuality and individual memory. It is argued that a medial turn within geography introduces heterogeneous time frames missing in traditional representations of space and place.

## Mediality and time

The assumption that new media introduce specific spatio-temporal frames for our perception of time draws on a line of thought pointing back at least to Walter Benjamin's (1969) famous essay, 'The work of art in the age of mechanical reproduction'. Benjamin points out that not only nature but also historical circumstances – including media development – organise and re-organise human perception. New modes of production can lead to new modes of cultural reproduction and introduce new modes of perception (Benjamin, 1969: 222). Benjamin sees the big shift in modernity in the possibility of mechanical reproduction of cultural artefacts that leads to a detachment from tradition and authorship. He uses the example of film to explain how the new medium introduced a new temporal structure leading to a shift in perception from concentration to '[r]eception in the state of distraction' (Benjamin, 1969: 240), well in line with the increased pace of mechanical industrial work and the modern city.[2]

Half a century later, Friedrich Kittler makes the notion of new media forming novel structures of space-time into an essential part of his media theory. In his workings, Kittler repeatedly shows how technical media alter the continuous flow of time. Above all, it is the media's constitutive ability to transform time into a spatial unit so as to make it manipulable that makes the difference; this function, according to Kittler, specifies the mediality of all technical media, from the gramophone over film to the universal Turing machine (Kittler, 1999). Sybille Krämer calls this the 'Cultural Techniques of Time Axis Manipulation', highlighting the significance of the media's own spatio-temporal structure for culture at large, and she argues that in Kittler's media theory, technology provides a means for channelling temporal irreversibility (Krämer, 2006).

Because the cinematic image has been perceived by media theorists as radically different in its representation of time, this chapter starts by highlighting the similarities and differences between the temporal structures of the moving image and geographic media. It continues by showing how the integration of time changed with the introduction of navigational means in geomedia, and then outlines the influence of algorithmic image production onto their temporal frame, using Deleuze's term of the 'crystal-image' (Deleuze, 1989: 89). In a final step, these heterogeneous temporalities are confronted with subjective perceptions of time with reference to Bergson's concept of duration (Bergson, 2005 [1988]). By doing this I highlight different interminglings between strategies of time integration in digital geomedia, foregrounding the relations between temporality and the technicity (Ash, 2012) of the artefact and its use. For example, contents are transmitted through the time stamp of images, or the time of origin suggested by the aesthetics, while the individual user's perception of time, memories and longing for the past are incorporated into the reading process. Therefore, when taking the actual usage into account, visual geomedia afford a *threefold temporality* where memories meet visual images which have been algorithmically stitched together, and get interconnected through navigation by the users.

Digital geomedia, such as Google Earth, serve as objects for investigating heterogeneous and artificial cultural and perceptual temporalities. I use this example of Google Earth as a 'theoretical console' (Verhoeff, 2009; 2012a). Verhoeff deploys this term to illustrate how the study of objects as indicators can point to the significance of contemporary theoretical conceptions as part of the discourse in culture and society. Following this methodological and theoretical position, geomedia applications become more than direct implementations of theoretical thought and instead interact with, and shape, theoretical conceptions of time and space. In this case, it is argued that the heterogeneous temporalities inscribed in geomedia are indications of a shift in the perception of cartographic space and time. Taking Google Earth as a telling example, I will argue that in contemporary networked and navigational geomedia, structures of space-time are not as clearly discernible as in older technical media but exhibit a stratified structure. This becomes clearer when we take a look at how contemporary interfaces of geomedia encapsulate different temporal frames within a single surface.

## Temporality by animation

The first mode of temporal integration in geographic media can be described as the convergence of animation and cartographic inscriptions. The essential prerequisite for this has been the disengagement of the image from paper as its carrier medium. In the beginnings of electronic cartography, there were experiments with different display technologies. For example, in 1966, Bruce Cornell and Arthur Robinson discussed a method to depict movement with the help of a computer attached to a cathode ray tube. In this process, a line-drawing ray of light was filmed by a 16-mm or 35-mm movie camera frame by frame. The results were animated maps that were used to depict meteorological data and the movements of satellites (Cornwell and Robinson, 1966: 79). The remediation of the cartographic sign plate, combined with the coupling of cathode ray tube, computer, input device (light pen) and film camera, added movement to the cartographic presentation. The film recording served as an intermediary to enable the publication and circulation of cartographic movement. Beginning in the 1950s, the cathode ray tube in the form of the television was the main source for the diffusion of animated maps, and one of the first regularly televised animated maps were weather maps (Cartwright, 2007: 14). But even before that, animation was widely used in so called cinematic maps and can be traced back to the first docudramas of the 1910s (Caquard, 2009).

But strictly speaking, it is wrong to call the animated maps used in films cinematic, because, as Deleuze points out, '[a]ny other system which reproduces movement through an order of exposures [*poses*] projected in such a way that they pass into one another, or are "transformed", is foreign to the cinema' (Deleuze, 1986: 5). The difference between cinematic and animation film lies therefore in the treatment of space by the medium: 'the cartoon film is related not to a Euclidean [*sic*], but to a Cartesian geometry. It does not give us a figure described in a unique moment, but the continuity of the movement which describes the figure' (Deleuze, 1986: 5). In animated visual media, time is on the whole subordinated to movement. So instead of regarding the cinematic map as the animation of a single sign plate, another perspective might be to regard it as a filmed geography, set in motion not by changes on the surface, but through the succession of discrete shots. Historically, the cathode ray tube was the medium of choice here as well. At the end of the 1970s the Massachusetts Institute of Technology (MIT) created hypermedia systems that promised to enhance and extend hypertext with video images, and thus enabled movement as a result of interactive navigation. One of these systems is a virtual car ride through the city of Aspen in Colorado. The Aspen Movie Map (1978–1979) is a hybrid of video display and navigational interface that has been described as an image-based surrogate travel system. At the lower half of a cathode ray tube are arrow-shaped touch-sensitive interface elements accompanied by a topographical minimap that depicts the user's current position, the next crossing and the route already travelled. After touching the on-screen arrows, the system selects the appropriate film sequence on a laser disc. While playing the video recording of the actual travel through the streets of Aspen the impression of real-time movement is simulated. The developer of the Aspen Movie Map, Andrew Lippman, calls this 'tour[ing] the area' (Lippman, 1980: 32).

This individualising of spatial on-screen movement at the MIT is significant for the further development of digital cartography, and signals a paradigm shift in the experience of moving geographic images: a shift from animated presentations to navigable geography-based applications. For Lev Manovich these early hyper media systems illustrate a fundamental principle of new media, which he labelled 'variability':

> in hypermedia, the multimedia elements making a document are connected through hyperlinks. Thus the elements and the structure are independent of each other – rather than hard-wired together, as in traditional media. (Manovich, 2001: 38)

The continuing convergence of cartography, film and computer technology results in an aesthetic transformation of time-cartography in motion. Along with this shift has come a renunciation of traditional static methods of geographic representation. At this juncture, I want to draw on a more recent example that seems symptomatic of the shift from static representations of space; via animated maps through to navigational geomedia. My exemplary theoretical object of choice here is the integration of Google Earth imagery in a Hollywood blockbuster, as illustrated in the following description:

> after a wild chase through a shopping mall Chev Chelios, protagonist in the action movie *Crank* (2006), jumps on the backseat of a waiting taxi. The driver, noticeably unimpressed, asks him straight away, 'Where do you wanna go, man?' Sensing the urgency of the situation, he barely waits for the answer ('Beverly Hills'), and hits the road with squealing tyres. The camera remains in one spot, showing the lateral view of the car in a slight low angle shot moving to the right, and at the moment when only the trunk is within the picture, the background changes to a satellite view of Los Angeles. The satellite imagery in top-down view is assembled with the perspective film image; the camera picks up the dynamic of the vehicle and follows it to the right until the taxi leaves the frame. Suddenly, it zooms out of the satellite image, describes a half circle until it zooms back into the picture, and finally comes to a stop above a pool on the roof of a mansion. 'Carlito's Penthouse' is written with red letters onto the image. After a short halt the camera zooms towards the pool for a few more frames. There

the framing changes within an abrupt transition from a perpendicular view back into lateral view from where the storyline is picked up again. (Author's own synopsis)

Whole sequences in the action movies *Crank* (2006) and *Crank 2 – High Voltage* (2009) are produced using the geobrowser Google Earth. What happens to the cartographic image and what happens with cinematic time here? At first glance the effect of the use of Google Earth in *Crank* is congruent with the general functions of maps in movies, which are: to claim the authenticity of a place, to serve as a narrative device in the form of cartographic intertitles, and to provide a jump effect, when a journey is traced on a map (Caquard, 2009). But there is more to it. In the case at hand the cartographic imagery is not just intertitle, or solely a jump effect. The geobrowser imagery itself becomes a motion picture technology. The sequence is inserted to establish a manipulation of the time axis through which cinematic time is itself accelerated. As a result, the time of travel is adjusted to the overall rhythm of the movie, which draws on the suggestion of real-time action. Time-space gets compressed while the cinematic image is replaced by a geographic surface image. The contraction of space allows for a compression of narrated and screen time.

This reverse remediation of Google Earth is not purely aesthetic (see Figure 4.1). The original interface items – the compass rose and the copyright of the image provider '© 2006 Sandborn' – are visible in the lower edge of the screen. The 'authorised' interface of Google Earth is transformed into a cinematic medium to traverse space. Integrated in the movie, the locomotion of the parabolic flight, in Google Earth a primary component of navigation, becomes an alternative to cuts and shutter dissolves and affords a cut-free traverse through the space between two scenes. In addition, the flight serves as a marker for a change of place and replaces the arrival normally signalled by an establishing shot. A new form of travelling with the virtual camera comes into existence by the montage of film images and satellite views. But the integration of the local context of the cinematic scene into the global context of the cartographic image and back also plays out the different aesthetics of the two types of images. Within the movie, the usage of satellite imagery points to the increasing capabilities of geosurveillance architectures that are able to track down and visually trace subjects through space and time and even predict their soon-to-be destinations.

This example tells us more about the mediatisation of cartography than about the integration of maps in movies. The key actor of this altered presentation and experience is the virtual camera that transforms the parameters of reception in computer-generated environments. The possibility to zoom in and freely choose the image section and the degree of mobility changes the subject–object

**Figure 4.1** Compilation of stills from the movie *Crank* (dir. Mark Neveldine and Brian Taylor/Lakeshore Entertainment, Lions Gate Films, RadicalMedia, GreeneStreet Films/ USA/2006)

relationship. In contrast to a movie camera, the virtual camera mobilises not the image but the gaze of the user. In Google Earth, the virtual camera and the spectator enter a novel constellation where competences are distributed anew. It is the framing of space that is affected by interactive navigation.

Traditionally, framing entails the preselection of the media space by the camera operator as an intermediator licensed to steer the attention of the passive spectator.[3] Thus in classical cinema, time as movement remains in the hands of the producers, while in interactive environments the spectator becomes the mobilised agent of change. This possibility of spatial transition with the virtual camera opens up a new media space that Mitchell Schwarzer (2004: 12) calls 'zoomscape'. In reference to Virilio's (1986: 8) 'dromology', the term refers to a space perceived in motion either mediated, or experienced, from within a transport vehicle. According to Schwarzer (2004), as far as the aesthetic effect is concerned, both modes can be put on the same level, which is why he proposes to substitute the term movement for the term mobility. As distinguished from movement, mobility not only captures the actual movement of the body but also the latent mobility of stationary perception of moving images, in addition to the simulation of movement while navigating an interface.

In environments that can be navigated by the user, determining framing and assembling constituent elements within a frame are obsolete, there is no particular camera perspective and we cannot speak any longer of a prescribed setting, trajectory or time frame. The totality of the environment becomes a potential focus for the user-spectator. Similarly, mobile geomedia turn the view of the built-in smartphone camera into an interface, so the 'actual' environment of the subject using the device becomes a potential target. The whole of the environment – whether virtually or geographically real – becomes a potential interface, a mode of production that corresponds to the conditions of production of digital cinema (Manovich, 1999) and of immersive virtual environments in general.

With navigation, any producer-sided predictability of framing and mobility is dropped. If the virtual environment of the geobrowser is compared to the cinematic universe, every visual aspect can be brought into focus. The functional shift of the camera eye can be termed a 'mise-en-space' (Jones, 2007: 227) in order to distinguish it from the traditional cinematic art of *mise-en-scène*. As a result of this shift, the camera evolves from being a creative element to a primal tool of creativity itself (Jones, 2007). With the interactive virtual camera, framing is no longer delegated to an authorised alliance of operator and apparatus. The operator-apparatus becomes a mediator by order of the spectator-user, who is now in charge of the perspective and content selection. The map is transformed into a space of action.

The divergent production of time is significant here. Through navigation, the temporality inherent to the medium coincides with the temporality of its use: the missing sequencing and montage leads to the situation in which only a chronology of actions exists. Flashbacks can come about only by reversion of time and repetition, not by cutting and jumping back. In comparison to the cinematic handling of time, the act of creation in digital geomedia happens in the real-time of use. As a sequel of these transformations the spectator-now-user becomes the creator of an individual spatial narration.

The Google Earth sequence in the movie *Crank* can be thought of as an example of the middle-ground between the animated and the navigational map, still bound to the cinematic time, but already mimicking its collapse with the introduction of the gaze of the virtual camera. So the development of dynamic mapping becomes a history consisting of the mobilisation of cartographic images and the map user's gaze, where different steps of animation and phases of user mobilisation can be identified. Both modes ensure that time can enter cartographic visualisations, first with animated maps, and then by means of mobilising the cartographic gaze. Through these developments, cartographic mobility ultimately approaches a navigational regime (Verhoeff, 2012a).

## Geobrowsing – temporality by navigation

To further grasp the distinction between static and dynamic geographic representations in theoretical terms, one can turn to the dichotomy of mimetic and a navigational map production and use. November, Camacho-Hübner, and Latour (2010) use this dichotomy in order to describe two different systems associated with the cartographic image. With regard to geomedia, they observe a major change in the epistemological foundation of cartography with the digital upheaval. The shift from isolated and static geographic representations to networked and interactive geomedia reconfigures the relationship between map and territory:

> while, in pre-computer times (B.C. [before computers], as geeks say), a map was a certain amount of folded paper you could look at from above or pinned down on some wall, today the experience we have of engaging with mapping is to log into some databank, which gathers information in *real time* through some interface (usually a computer). (November, Camacho-Hübner and Latour, 2010: 583)

New geographic media alter our perception of maps insofar as a mimetic interpretation impedes a navigational interpretation of maps. While in pre-digital times people looked at maps from above, in the age of digital cartography, map reading means to 'log into some databank, which gathers information in real-time through some interface (usually a computer)' (November, Camacho-Hübner and Latour, 2010: 583). From these observations, the authors deduce a theory about the changing referential status of maps based on the epistemological gap between map and territory or between signifier and signified (Korzybski, 1994 [1933]) that is negotiated step by step. They suggest that navigational interpretation is strongly facilitated in digital mapping. Thus, mimetic resemblance is a necessity but not necessarily a sufficiency to get from territory to map and vice versa. Following William James, the accuracy of the representation is of less importance than its ability to lead through successive signposts (James, 1996 [1907]). Thus, the navigational paradigm relativises indexicality until a 'circulating reference' (Latour, 1999: 24) becomes visible which enables varying degrees of resemblance of the signifier with the signified in an action space. The navigational approach assumes that map use in the digital age can be modelled as a trajectory along a chain of inscriptions, where each image is matched to the previous and the succeeding one rather than to some external point of reference.

This distinction adds to the performative critique invoked in the introduction, which questions the status of the map as a static representation of a territory. Massey (2005: 107) suggests that 'the dominant form of mapping, though, does position the observer, themselves unobserved, outside and above the object of the gaze'. Massey of course also acknowledges that there are static maps that tell stories, for example the Christian *mappae mundie* like the Ebstorf Map, but the story is usually framed within a closed system – in the example, pagan and biblical history epistemologically and graphically encloses the map's content within a circular space around the body of Christ.

The navigational, by contrast, emphasises the performative character of the mapping as an open process, and at the same time, the involvement of the spectator is thus able to challenge static and linear models of cartographic communication that are based on unified and standardised ontologies. Turnbull (2007), for example, criticises 'Western' mapping practices for dismissing the local and navigational dimension of spatio-temporal knowledge. There is an incommensurability between traditional, local (indigenous) knowledge and (Western) knowledge traditions of science because indigenous knowledge gets translated into a unified ontology by deflation:

> mapping and databasing using scientific coordination of commensurability techniques not only subsume differing spatialities and temporalities into one abstract space-time they also omit the multiplicitous and interactive dimensions of the local and the practical, the stories and the journeys, the spiritual and the experiential. (Turnbull, 2007: 141)

Here performative critique is expressed in a concern about scientific knowledge becoming naturalised, when any trace of the local human journey or passage disappears. Eventually, this impetus to re-model cartographic communication with dynamic, performative and mobile parameters points to a general critique on the scientific self-conception and on the claim to truth of scientific facts. Following

Bergson, this distinction between mimetic and navigational also reflects a difference between the scientific use of images and the world of consciousness:

> now no philosophical doctrine denies that the same images can enter at the same time into two distinct systems, one belonging to *science*, wherein each image, related only to itself, possesses an absolute value; and the other, the world of *consciousness*, wherein all the images depend on a central image, our body, the variations of which they follow. The question raised between realism and idealism then becomes quite clear: what are the relations which these two systems of images maintain with each other? And it is easy to see that subjective idealism consists in deriving the first system from the second, materialistic realism in deriving the second from the first. (Bergson, 2005 [1988]: 26)

If we follow Bergson even further, the line between these two positions can be drawn by looking at the different conceptions of the temporal structures of images and their relation to the body. The translation of images into actions by the mind is analogous to the transformation of a mimetic image into a navigational dynamic. In traditional map production, the mimetic is preferred over the dynamic in order to preserve, stabilise and mobilise geographic knowledge; a panoptic vision of the world is propagated. However, the navigational advocates a mode of spatial perception that can be thought of as an act of performativity, involving the interplay between vision and movement (Gell, 1985), or between image capture and experience (Verhoeff, 2012b).

Taken as a theoretical object, Digital Earth, as outlined by Al Gore in the 1990s, has indeed promoted a mimetic, holistic and all-encompassing world view. Thus, versions of Digital Earth could become tools of what Donna Haraway calls 'the god trick' (Haraway, 1988: 582). The product management of Google Earth also follows this process, but with the twist that Google introduces a vertical continuum, designed to cover all zoom levels, from the so-called ionic view of under one meter, to the arm's length away Google Goggles, through the panoramatic Street View to the Google Map view, and even further up to the orbital distance of Google Earth. The difference between Google Earth and the perfect panopticon is the lack of real-time imagery. A genuine panoramatic 'scopic regime' (Knorr-Cetina, 2009: 64) can only be effective if it is able to register and display every change in the status quo. But the seamless navigation in geomedia, along with the smoothing of zoom levels, only tricks us into the feeling of panoptic visibility, covering up the layered imagery, and the juxtaposition of different media with varying time stamps. The surface of the map strives for a geovisual holism. Thus, the mediation evokes the illusion of objectivity, totality and transparency whereas in reality, such a comprehensive view is merely illusory. At this point Latour draws our attention to the manifold of 'oligopticons' (Latour, 2005: 175) that counteract conceptions of hierarchical power structures as determinants of human agency. According to Latour, the oligopticons are little windows which only allow a snipped view of the social world when taken in isolation. Oligopticons point to the deflation strategy mentioned in the introduction, since all facts are flat in the sense that understanding the world is not a matter of scale but a matter of connecting the dots and following the associations among them.

If we take Google Earth as a theoretical object, it helps us to picture the shift between the all-encompassing idea of one panopticon, to a relational understanding of many interconnected oligopticons. The former is suggested by the overall presentation of the whole, projected onto the supratemporal 'WGS84' geodesic reference system; the latter is the image detail selected by the virtual camera cut-out while time-critically navigating the interface. The fact that the imagery of the geobrowser's basic layer is made up of many images, each taken and produced at a different time, means the heterogeneous time frame of the geobrowser alone obscures the panoptic order. One could state that Google Earth supports heterogeneous temporalities within a single scopic regime through the juxtaposition of images with diverging time stamps, and by offering the means to virtually navigate them. In comparison to other panoptic regimes, Google Earth pre-structures the user experience of time-space in an incongruent fashion. This possibility to pick your own oligopticons brings about new possibilities of being surprised by the encounter with cartographic space (see Massey, 2005: 116):

> we pan the map, zoom in and zoom out, and change colors. All of these involve 'playing' with the map to allow latent relationships to emerge. There are other ways of manipulating maps for this purpose that we may not ordinarily do – turning the map upside down and sideways, for example. (Peuquet and Kraak, 2002: 82)

In geobrowsing, a playful cartography is acted out that is closer to a navigational form of cartography than to the mimetic paradigm. The map-supported appropriation means transforming the abstract logocentric space shown in a geobrowser into a personal egocentric space through interactive adjustments to the position and perspective of the viewer's visual focus. The movement involved here is a form of knowledge generation since it allows for 'latent relationships to emerge' between different oligopticons, while the ever changing surface of the geobrowser bears the possibility to show that 'all spaces are, at least a little, accidental, and all have an element of heterotopia' (Massey, 2005: 116). Geobrowsing allows for this kind of emergence in an otherwise premodelled environment, and emergence is only possible by linking perception to movement and as such by letting time lapse away.

## Algorithmic temporalities

A third temporal dimension is closely associated to the technicity of the geomedia interface and therefore gets inscribed during the production process. Uricchio speaks of a spatial heterogeneity of digital media that comes from the technique of stitching images of different perspectives together (Uricchio, 2011). He draws on applications like Microsoft's Photosynth that combine geotagged photographs taken by different individuals to form navigational mosaics of digital images. The images have been taken from different angles, cameras, standpoints and with the help of different vehicles, like cars or satellites. They get algorithmically combined – stitched together – through calculation of overlaps and junctions in order to form a coherent whole. With enough data, three-dimensional spaces emerge which allow the user seamless navigation. In order to analyse the status and value of these image spaces, it is not sufficient to have the production process on one side and the user on the other, without showing an interest on the mediating entity in between. But Uricchio does not lament the loss of reference of digital images. His posture is not an ontological one. Instead, he takes a closer look at the aggregation and combination of images by means of 'algorithmic interventions' (Uricchio, 2011: 25) which lead to a reconfiguration of subject–object relations. Eventually algorithmic interventions circumvent conventions such as linear perspective and challenge notions of authorship.

For example, single images of Google Earth are taken in different seasons and in different weather conditions. Resultantly, it can happen that while geobrowsing, the sun is shining and the trees are carrying green leaves, and within one second, it is getting dark and the trees show empty branches: seasons change by half-turns (see Figure 4.2). Thus, geobrowsing happens within an ambiguous temporality that is archetypal for the distributed aesthetics of new media (Lovink, 2008: 225). The heterogeneous temporalities of video platforms like YouTube come to mind where videos and commentaries of different dates of origin are combined, the timelessness of circulating memes that surface time and again, or the combination of contents on a Facebook profile site that the personal timeline organises in an only artificial chronology.

Avant-garde adoptions of geomedia, as also discussed in other chapters of this book, make use of algorithmic interventions to expose the delicacy of the boundaries among multiple spatio-temporal standpoints. Artists use geographic imagery as raw materials, which they recombine, modify or alienate in order to distill their specific medial engagements. One example is Christoph Engel's image series *Sommet*, for which the artist took screenshots of three-dimensional textures of mountains in Google Earth and altered the imagery with tonal

**Figure 4.2** Seasons change by half-turns in Brooklyn, New York. Google Earth images © Google. This figure has not been made available under a CC licence. Permission to reproduce it must be sought from the copyright holder.

correction until they resemble analogue nineteenth-century travel photography (Engel and Westphal, 2009). Engel poses the question of authenticity since in these images the production date becomes an illusion, only maintained as long as the spectator keeps their distance. From close up, the pixelated mediality of the three-dimensional models shows up. But in a previous step, the treatment of the artist has already transformed the textures and satellite imagery into travel media; moving them away from their purpose as means of land surveying, map production and remote sensing through falsification of the time of creation. Another artist, Clement Valla, looks for already inscribed distortions, anomalies and non-standards in geomedia:

> at first, I thought they were glitches, or errors in the algorithm, but looking closer, I realized the situation was actually more interesting – these images are not glitches. [...] These jarring moments expose how Google Earth works, focusing our attention on the software. They are seams which reveal a new model of seeing and of representing our world – as dynamic, ever-changing data from a myriad of different sources – endlessly combined, constantly updated, creating a seamless illusion. (Valla, 2012: no pagination)

Valla points to the specific mediality of Google Earth as a space of action that renders visible the polysemantic value of geomedia and the user's role in the process of decoding these artefacts by exposing 'algorithmic agency'. The glitch as the most visible evidence for an algorithmic agency at work becomes an element which is tied closer to the mediality of geomedia, and therefore exposes and uncovers part of the production process as well as allowing us a glimpse of the heterogeneous spatiality. While this notion of algorithmic intervention is focused on changes in the visual appearance of media and points to the impact automatic production has on the conception of space, there is an underlying time-critical aspect to it as well. What I want to stress here is the heterogeneous time frame introduced though applications that stitch together imagery from various sources, but not as something that prevents enjoyment or proper use. Instead, this uncertain temporal structure of the images in geomedia seems to invite a playful encounter with the aesthetics that are freed from a particular historicity.

The artistic examples above also exemplify how the context and framing of images and activities can be more important than indexicality in the sense of mimetic correspondence between map and territory. Instead, they show that referential cartographic indexicality can be replaced by polysemantic values in geographic images, which allow alternative interpretations without relying on a full correspondence to the depicted scenery, building or landscape. Thus glitches also point to heterogeneous times which intermingle within one image in the interface of geomedia. Deleuze uses the term of the 'crystal-image' (Deleuze, 1989: 89) to describe the simultaneous existence of the past, the present and ultimately the future within one image. He links the 'crystal-image' to the affordance of film that mingles different temporal horizons – usually done by addressing different 'sheets of the past' (Deleuze, 1989: 98) simultaneously with the help of mirrors or montage. Comparably the glitch becomes an image that points back to the bygone process of its own making. Thus, the glitch is part of the navigational paradigm of geomedia which integrates isolated cartographic images and re-connects them to the visualisation as a whole.

## Subjective time and individual duration

We cannot help but transform geographic space and its abstract depictions into spaces filled with our own memories and recollections of the past. Therefore, subjective perception and individual experience of time adds an additional temporal layer to geomedia applications. This final temporality at work is concerned with the subjective time of memory and remembrance of the locations of past events and the places along individual biographies. At this point we can draw on Bergson's (2005 [1988]) concept of 'duration' (*durée*) in order to describe the temporality of the experience with geomedia. Bergson (2005 [1988]) describes duration as the qualitative aspect of time, distinct from the quantitative concept of linear time that can be segmented in consistent intervals. '[T]here is no perception which is not full of memories. With the immediate and present data of our senses, we mingle a thousand details out of our past experience' (Bergson, 2005 [1988]: 33). As a consequence, while looking at it, while navigating through it, and while getting immersed in it, the abstract space of the map gets inevitably transformed into a personal space of remembrance. This makes geomedia suitable for evoking a feeling of nostalgia which is, at first sight, a structure of feeling that is inherently location-specific, but in fact it is a time-related yearning: 'at first glance, nostalgia is a longing for a place, but actually it is a yearning for a different time – the time of childhood, the slower rhythms of our dreams' (Boym, 2001: xv). Seen this way, the return to a place of the past – either physically or virtually – is the attempt to turn back time (Boym, 2001: xv–xvi). At this point, the time of the images, the narrative time of the medium so to speak, meets individual time; an encounter mediated by the navigational options of the interface. The retrieval of the places of the past depends on the congruence between the time of origin of the imagery, which as we have seen is heterogeneous in itself.

## The heterogeneous temporalities of geomedia

For Bergson, the present is only a mixture and every recollection is nurtured by superimposed strata of the past (Bergson, 2005 [1988]). Deleuze writes about this organising principle of remembrance:

> depending on the nature of the recollection that we are looking for, we have to jump into a particular circle. It is true that these regions (my childhood, my adolescence, my adult life, etc.), appear to succeed each other. But they succeed each other only from the point of view of former presents which marked the limit of each of them. (Deleuze, 1989: 99)

Put into practice, within geomedia this process becomes a feedback loop which constantly oscillates between the past and the past's presents: 'we have to jump into a chosen region, even if we have to return to the present in order to make another jump, if the recollection sought for gives no response and does not realize itself in a recollection-image' (Deleuze, 1989: 99). Similarly, the process of transforming the abstract and distanced space depicted in geomedia into one's own space is acted out by taking the trip down memory lane. Time and space become subjective in the course of this, but not in the way that we internalise it. Our share of time also becomes visible and spatialised. As Deleuze (1989) emphasises, in Bergson's thought time is never subjective in the sense that it becomes interior to us. Instead, time is the interior in which we exist and consequently we can solely make the past become an interior *image* of us. According to this, visual geomedia, like geobrowsers, can serve as a medium of remembrance that translates their own biography into a topology of events. But only in navigating this trajectory can we overcome the stasis of this representation and start to virtually travel through our own past.

This results in a co-existence of the past and the present within a composite that is actualised in the act of navigation. Thus, geobrowsing, as the practice of using digital cartographic interfaces, turns into a game of *threefold temporality*: individual duration meets the heterogeneous time frame made of algorithmic imagery, which is then navigated by the users. So, on the one hand, contemporary geomedia act like traditional maps as agents of a spatialisation of time, supporting the tendency to convert time into space relations (Gross, 1981). However, on the other hand, a medial turn introduced by geomedia brings back heterogeneous time and individual duration that have been missed out in traditional, static geographic inscriptions. Within this process, representations of space and place that formerly relied on strictly standardised ontologies become augmented with subjective and local knowledge. Therefore, dynamic, performative and playful encounters with geographic space are the result of heterogeneous temporalities that challenge the immutability of the sign plate. With maps becoming 'navigational interfaces' (Lammes, 2011: 1), alternative temporalities enter cartographic interfaces and representations become more mutable, as cartographic indexicality is not subordinated to an absolute time frame but becomes just another layer of experience.

## Notes

1 Tristan Thielmann's definition encompasses locative media and mediated localities (see Thielmann, 2010).
2 'Our taverns and our metropolitan streets, our offices and furnished rooms, our railroad stations and our factories appeared to have us locked up hopelessly. Then came the film and burst this prison-world asunder by the dynamite of the tenth of a second, so that now, in the midst of its far-flung ruins and debris, we calmly and adventurously go traveling. With the close-up, space expands; with slow motion, movement is extended. The enlargement of a snapshot does not simply render more precise what in any case was visible, though unclear: it reveals entirely new structural formations of the subject' (Benjamin, 1969: 236).

3 'We will call the determination of a closed system, a relatively closed system which includes everything which is present in the image – sets, characters and props – *framing*. The frame therefore forms a set which has a great number of parts, that is of elements, which themselves form sub-sets. It can be broken down' (Deleuze, 1986: 12, original emphasis).

## References

Ash, J. (2012) Technology, technicity, and emerging practices of temporal sensitivity in videogames. *Environment and Planning A*, 44(1): pp. 187–203.

Benjamin, W. (1969) 'The work of art in the age of mechanical reproduction'. In: Arendt, H. (ed.) *Illuminations*. New York: Schocken Books, pp. 217–251.

Bergson, H. (2005 [1988]) *Matter and Memory* (8th edn). New York: Zone Books.

Boym, S. (2001) *The Future of Nostalgia*. New York: Basic Books.

Caquard, S. (2009) Foreshadowing contemporary digital cartography: A historical review of cinematic maps in films. *The Cartographic Journal*, 46(1): pp. 46–55.

Cartwright, W. (2007) 'Development of multimedia'. In: Cartwright, W. (ed.) *Multimedia Cartography*. Berlin: Springer, pp. 11–34.

Cornwell, B. and Robinson, A. H. (1966) Possibilities for computer animated films in cartography. *The Cartographic Journal*, 3(2): pp. 79–82.

Deleuze, G. (1986) *Cinema 1: The Movement-Image*. Minneapolis, Minnesota: University of Minnesota Press.

Deleuze, G. (1989) *Cinema 2: The Time-Image*. Minneapolis, Minnesota: University of Minnesota Press.

Döring, J. and Thielmann, T. (2009) *Mediengeographie: Theorie – Analyse – Diskussion*. Bielefeld: transcript Verlag.

Engel, C. and Westphal, U. (2009) *Ungefähre Landschaft/Approximate Landscape*. Berlin: Deutscher Kunstverlag.

Felgenhauer, T. and Quade, D. (2012) 'Society and geomedia: Some reflections from a social theory perspective'. In: Jekel, T., Car, A., Strobl, J. and Griesebner, G. (eds) *GI_Forum 2012: Geovisualization, Society and Learning*. Berlin: Wichmann, pp. 74–82.

Gell, A. (1985) How to read a map: Remarks on the practical logic of navigation. *Man*, 20 (2): pp. 271–286.

Gross, D. (1981) Space, time, and modern culture. *Telos*, 50: pp. 59–78.

Haraway, D. (1988) Situated knowledges: The science question in feminism and the privilege of partial perspective. *Feminist Studies*, 14(3): pp. 575–599.

Ingold, T. (2009) 'Against space: Place, movement, knowledge'. In: Kirby, P. W. (ed.) *Boundless Worlds: An Anthropological Approach to Movement*. Oxford: Berghahn Books, pp. 29–43.

James, W. (1996 [1907]) *Essays in Radical Empiricism*. London: University of Nebraska Press.

Jones, M. (2007) Vanishing point: Spatial composition and the virtual camera. *Animation* 2: pp. 225–243.
Kittler, F. (1999) *Gramophone, Film, Typewriter*. Stanford, California: Stanford University Press.
Knorr-Cetina, K. (2009) The synthetic situation: Interactionism for a global world. *Symbolic Interaction*, 32(1): pp. 61–97.
Korzybski, A. (1994 [1933]) *Science and Sanity. An Introduction To Non-Aristotelian Systems and General Semantics* (5th edn). New York: Institute of General Semantics.
Krämer, S. (2006) The cultural techniques of time axis manipulation: On Friedrich Kittler's conception of media. *Theory, Culture and Society*, 23(7–8): pp. 93–109.
Lammes, S. (2011) 'The map as playground: Location-based games as cartographical practices'. In: *Think, Design, Play: Proceedings of the Fifth International DIGRA Conference*. Utrecht, pp. 1–10.
Latour, B. (1986) 'Visualization and cognition: Thinking with eyes and hands'. In: Kuklick, H. (ed.) *Knowledge and Society: Studies in the Sociology of Culture Past and Present Volume 6*. New York: Jai Press, pp. 1–40.
Latour, B. (1987) *Science in Action. How to Follow Scientists and Engineers through Society*. Cambridge, Massachusetts: Harvard University Press.
Latour, B. (1999) *Pandora's Hope: Essays on the Reality of Science Studies*. Cambridge, Massachusetts: Harvard University Press.
Latour, B. (2005) *Reassembling the Social. An Introduction to Actor-Network-Theory*. Oxford: Oxford University Press.
Latour, B. (2009) 'Die Logistik der immutable mobiles'. In: Döring, J. and Tristan, T. (eds) *Mediengeographie*. Bielefeld: transcript Verlag, pp. 111–144.
Lippman, A. (1980) 'Movie-maps: An application of the optical videodisc to computer graphics'. In: *Proceedings of the 7th Annual Conference on Computer Graphics and Interactive Techniques*, Seattle, pp. 32–42.
Lovink, G. (2008) *Zero Comments: Blogging and Critical Internet Culture*. New York: Routledge.
Manovich, L. (1999) 'What is digital cinema?' In: Lunenfeld, P. (ed.) *The Digital Dialectic*. Cambridge, Massachusetts: The MIT Press, pp. 172–192.
Manovich, L. (2001) *The Language of New Media*. Cambridge, Massachusetts: The MIT Press.
Massey, D. (2005) *For Space*. London: Sage Publications.
November, V., Camacho-Hübner, E. and Latour, B. (2010) Entering a risky territory: Space in the age of digital navigation. *Environment and Planning D: Society and Space*, 28(4): pp. 581–599.
Peuquet, D. and Kraak, J. M. (2002) Geobrowsing: Creative thinking and knowledge discovery using geographic visualization. *Information Visualization*, 1(1): pp. 80–91.
Rosenberg, D. and Grafton, A. (2010) *Cartographies of Time: A History of the Timeline*. New York: Princeton Architectural Press.
Schwarzer, M. (2004) *Zoomscape: Architecture in Motion and Media*. New York: Princeton Architectural Press.
Thielmann, T. (2010) Locative media and mediated localities: An introduction to media geography. *Aether*, 5a: pp. 1–17.

Thrift, N. (1977) Time and theory in human geography, part 1. *Progress in Human Geography*, 1: pp. 65–101.

Turnbull, D. (2007) Maps narratives and trails: Performativity, hodology and distributed knowledges in complex adaptive systems – an approach to emergent mapping. *Geographical Research*, 45(2): pp. 140–149.

Uricchio, W. (2011) The algorithmic turn, photosynth, augmented reality and the changing implications of the image. *Visual Studies*, 26(1): pp. 25–35.

Valla, C. (2012) The universal texture. Rhizome. [Online] Available at: http://rhizome.org/editorial/2012/jul/31/universal-texture/ (accessed 3 March 2018).

Verhoeff, N. (2009) Theoretical consoles: Concepts for gadget analysis. *Journal of Visual Culture*, 8(3): pp. 279–298.

Verhoeff, N. (2012a) *Mobile Screens: The Visual Regime of Navigation*. Amsterdam: Amsterdam University Press.

Verhoeff, N. (2012b) 'A logic of layers: Indexicality of iPhone navigation in augmented reality'. In: Burgess, J., Hjort, L. and Richardson, I. (eds) *Studying Mobile Media: Cultural Technologies, Mobile Communications, and the iPhone*. New York: Routledge, pp. 118–132.

Virilio, P. (1986) *Speed and Politics: An Essay on 'Dromology'*. New York: Colombia University.

# Part II

Stitching memories

# 5

## 'Space-crossed time': digital photography and cartography in Wolfgang Weileder's *Atlas*[1]

*Rachel Wells*

> The places we have known do not belong only to the world of space on which we map them for our own convenience. They were only a thin slice, held between the contiguous impressions that composed our life at that time; the memory of a particular image is but regret for a particular moment; and houses, roads, avenues are as fugitive, alas, as the years. (Proust, 2002: 513)

The creation of an 'Atlas' is an ambitious project. The word suggests accuracy in detail and comprehensiveness in scope. It suggests a certain degree of objectivity, despite the recent revolution in cartographic historiography that has emphasised, first, the role that maps play as instruments of power, second, their reliance upon their creator's particularity and creativity, and third, the need to examine maps within the broader cultural context of their creation and use (Cosgrove, 2008: 156). In 2011 the artist Wolfgang Weileder embarked upon his own *atlas-project*, an artwork which, I argue, reveals much about our relation to contemporary photography and its use online, particularly with regard to digital maps. Certainly Weileder's *Atlas* sits in line with Cosgrove's three observations about recent cartographic research: in making his own *Atlas* the artist implies the power that he yields, never attempts to hide his own authorship, and readily invites a detailed cultural and contextual interpretation. This chapter is an endeavour to begin the latter, by considering Weileder's *Atlas* within both his wider oeuvre and its positioning within contemporary photography and cartography. I suggest that at its heart, Weileder's *Atlas* reinforces a Proustian emphasis on the 'slicing' of time and memory across spatial referents, and that the artist's 'constructive' photographic practice – to use Walter Benjamin's term – suggests that

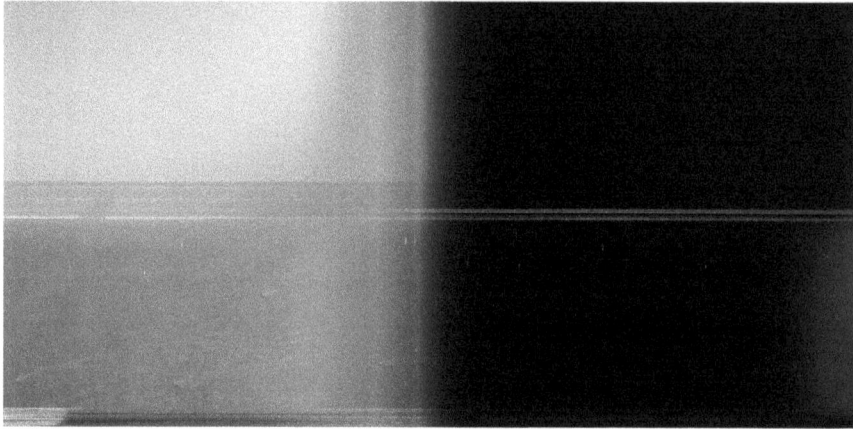

**Figure 5.1** Wolfgang Weileder, *Gulf of Naples s2912*, 2009, from the *Seascapes* series. Lambda Print, 75 x 154cm © Wolfgang Weileder (courtesy of Wolfgang Weileder). This figure has not been made available under a CC licence. Permission to reproduce it must be sought from the copyright holder.

the contemporary capitalist culture of the instant image is producing a form of illiteracy in experiencing and understanding the nexus between time and space. Maps and digital maps, however Dionysian in character according to geographers Kingsbury and Jones III, can fail to capture this Benjaminian sense of 'space-crossed time'; Weileder, like Proust, uses art to highlight the coagulation of fugitive years and roads, moments and avenues within human experience.

## Space-crossed time

In Wolfgang Weileder's *Seascapes* (2009–ongoing, see Figure 5.1), time marches steadily on, slice by slice, from left to right. These serene sunset images also function as graphs or 'waterfall displays', with time as an unremitting axis along the horizontal plane. Weileder has imposed a secondary rigorous system of recording to that of the photograph, so that the image is constructed as much as captured. The artist takes multiple digital images of the same scene at regular time intervals, and then extracts a one pixel-wide vertical strip from the same point in each photograph, laying the strips out in a row so that the composite image can literally be 'read' as a record of day turning into night. While in no way manipulating the initial image caught by the camera, Weileder's method involves a careful extraction from it, and a subsequent adding together and re-visioning of the photographic records. In doing so, the artist builds a new and abstract composite image. Rather than the orangey haze of conventional sunset images then, which focus on the nostalgic beauty of a sun soon to disappear, Weileder's *Seascapes* focus upon the effects of the presence or absence of a light source. Rather than photographing, in typical elegiac manner, the glowing transience of a closing day, Weileder's systematic recording across time produces rather than merely preserves the form of the image. The *Seascapes* present a different kind of beauty, with the symmetry of the image marked by a hazy, perfect cross; signalling the intersection of sea, sky, day and night. The coordinates of space and time are overlaid.

Without knowledge of Weileder's method, the *Seascapes* could be read very differently: the blurred edges of the temporal turning point between light and darkness could suggest the phosphorescence of a more physical planetary edge, as if the image had been taken from outer space, and the line between day and night marks instead the boundary between the inhabited world and the blank darkness of space. Reminiscent of Wittgenstein's famous 'duck-rabbit', the *Seascapes* images can transform for the viewer so that the temporal pivot becomes a spatial one as edges in time are read as shifts in space.

The conflation of time-space coordinates is a recurring element in Weileder's oeuvre. Photography's ability to act as a trace of time, to preserve a fleeting configuration of light and shadow, rarely seems enough in Wolfgang Weileder's varied practice. The conventional 'stilling' of photography is not sufficient to convey the intersection of time and space as conceived in his projects, whether photographic or sculptural. This implied dissatisfaction with the preservation enacted by photography resonates with the work of Hiroshi Sugimoto, whose own *Seascapes* (1980–2002) are referred to by Weileder's titling of his series with the same name (Weileder, 2013). Sugimoto's calm, atmospheric black and white photographs of various seas around the world are unified by sharing a consistent horizon line across each image, as if suggesting the immutability of the sea across both space and time (see Figure 5.2). Indeed, Sugimoto has discussed his *Seascapes* as an attempt to photograph that which does not change:

> I asked myself, 'can someone today view a scene just as primitive man might have?' The images that came to mind were of Mount Fuji and the Nachi Waterfall in ages past. A hundred thousand or a million years ago, would Mount Fuji have looked so very different than it does today? [...] Unfortunately, the topography has changed. Although the land is forever changing its form, the sea, I thought, is immutable. Thus began my travels back through time to the ancient seas of the world. (Sugimoto, 2010: 109)

Such a search for the unchanging marks Sugimoto's oeuvre, and highlights a curious desire to point the camera, a tool capable of preserving the fleeting instant,

**Figure 5.2** Hiroshi Sugimoto, *Seascape: North Atlantic Ocean, Cape Breton*, 1996 © Hiroshi Sugimoto (courtesy of Pace Gallery). This figure has not been made available under a CC licence. Permission to reproduce it must be sought from the copyright holder.

onto the 'immutable'. It is the reversed relationship between the transient and the permanent in Sugimoto's *Portraits* series (1999) that produces the eerie quality of the images. Photographs of Madame Tussaud waxwork figures preserve a fleeting view of the unchanging 'bodies', so that photography's oft-discussed relationship to death appears inverted, as if the camera's flash enlivened rather than entombed the historical figures already rendered resistant to ageing and decay (see Figure 5.3). In both series, Sugimoto suggests a desire to go beyond the preservation offered by photography, as if, as with Weileder's work, such preservation does not offer enough.

While Sugimoto's work indicates a dissatisfaction with pointing the camera at fleeting life, Weileder's work suggests a different dissatisfaction. Consistent within Weileder's use of photography is the implication that, to use Benjamin's phrase describing the aura of the photograph, its complex 'weave of space and time' is not – or perhaps is no longer – accurate enough (Benjamin, 1999a: 518). This limitation of the photograph is implied to be not only the result

'Space-crossed time'                                              119

**Figure 5.3** Hiroshi Sugimoto, *Henry VIII*, 1999, from the *Portraits* series. Gelatine silver print, 149.2 x 119.4cm © Hiroshi Sugimoto (courtesy of Pace Gallery). This figure has not been made available under a CC licence. Permission to reproduce it must be sought from the copyright holder.

of a withering away of aura at the hands of reproducibility, however. I will argue here that Weileder's projects suggest a dissatisfaction with the time-space coordinates of photography because of the speed, the instantaneous flashes, of our encounters with it. Weileder's *Seascapes* series is both a development of the ongoing *atlas-project* and a turning of it on its head; created using the same method of collating pixel-wide strips, the *Atlas* series focuses upon the activity of public places, with a vertical, rather than horizontal time axis. The resulting abstract images are therefore also maps measuring a particular intersection of geography and temporality (see Figure 5.4). Weileder has emphasised this reading of the *atlas-project* as a form of mapping through both the work's title and his own discussion of the creative process involved, stating that 'my photographic technology tries to combine both a cartographic way of seeing and a photographic representation over a long timescale' (Weileder, 2013: 93). Having emphasised our knowledge that 'everything is in a permanent state of flux', with reference to both theoretical physics and the economic climate, Weileder implies that a map delineating only the static aspects of a particular location will no longer suffice (Weileder, 2013: 93). The *atlas-project* suggests that both conventional photography and conventional cartography are essentially inadequate.

**Figure 5.4** Wolfgang Weileder, *Piazza San Pietro, Rome, Slice 1756*, 2010, from the *Atlas* series, 2011–present. Archival inkjet print, 140 x 22cm © Wolfgang Weileder (courtesy of Wolfgang Weileder). This figure has not been made available under a CC licence. Permission to reproduce it must be sought from the copyright holder.

Unlike Sugimoto's work, Weileder presents a consistent interest in the bustling change and fleeting presence of human temporality. His digital method, however, suggests that the conventional photograph does not fully capture the integration of this transience with space itself. Each 'map' within the *Atlas* series is accompanied by a key detailing the exact geographical location of Weileder's camera, along with information about what was happening in front of the lens, however extraordinary or mundane (see Figure 5.5). The artist has positioned his camera and stayed with it for an extended period of time in order to extract and then add together the 'slices' of everyday life that unfurled in front of him. It is possible to trace the movements of buses or of crowds across both place and time, so that the image is not a record of one instant, but a 'waterfall display' of space across time.

The implication that the photograph's work on its own is in some way insufficient is also present in Weileder's earlier work. If the *Atlas* and *Seascapes* series present a form of digital addition to the photograph's traces, then the photographs that form part of the *Transfer* (2006) and the *House* projects (2002–2004) present a similar desire to extend photography's temporal reach, but in analogue terms. As a trace of a transient process, these photographs were made using a very long exposure – sometimes ten days – so that the construction and deconstruction of Weileder's building projects is rendered as a ghostly half-presence

**Figure 5.5** Wolfgang Weileder, *Atlas*, exhibition view, Northern Gallery for Contemporary Art, 2013 © Wolfgang Weileder (courtesy of Wolfgang Weileder). This figure has not been made available under a CC licence. Permission to reproduce it must be sought from the copyright holder.

against the more definite blacks and whites of the surrounding environment (see Figure 5.6). As well as sharing the artist's concern with architecture and public place, both the *Transfer* and *Atlas* series reveal a motivation to extend photography's splicing of space and time: both space over time, and time over space. The photograph emerges in these artworks as a tool which must itself be constructed or deconstructed or both, in order to locate a sense of duration.

Given that the temporality of the photograph has long been a compelling source of intrigue for the medium's commentators, Weileder's desire to develop further its relationship between time and space perhaps reflects a wider shift in the relationship between the two that has been identified by current writers including David Harvey and Paul Virilio. In the 1930s, Walter Benjamin famously suggested that all timescales strangely coalesce within the photograph:

> no matter how artful the photographer, no matter how carefully posed his subject, the beholder feels an irresistible urge to search such a picture for the tiny spark of contingency, of the here and now, with which reality has (so to speak) seared the subject, to find the inconspicuous spot where in the immediacy of that long-forgotten moment the future nests so eloquently that we, looking back, may rediscover it. (Benjamin, 1999a: 510)

**Figure 5.6** Wolfgang Weileder, *House-Madrid*, 2004. Gelatine silver print, 79 x 100cm © Wolfgang Weileder (courtesy of Wolfgang Weileder). This figure has not been made available under a CC licence. Permission to reproduce it must be sought from the copyright holder.

The trace of the photograph, its indexical nature as described by Susan Sontag's analogies of a 'death mask' or 'footprint', presents viewers with a mysterious conflation of past, present and future: we seek out the 'now' of a past moment in order to *re*discover a previous future (Sontag, 1979: 154). Benjamin's delight in such a magical overcoming of time's linearity is evident in his discussion of David Octavius Hill and Robert Adamson's photograph of Mrs Elizabeth Hall:

> with photography, however, [in contrast to painting] we encounter something new and strange: in Hill's Newhaven fishwife [...] there remains something that cannot be silenced, that fills you with an unruly desire to know what her name was, the woman who was alive there, who even now is still real and will never consent to be wholly absorbed in 'art'. (Benjamin, 1999a: 510)

Even if accepting Mary Price's criticism that Benjamin 'has a talent for characterising a still photograph as a narrative, implying the beginning and end of a situation by his dramatic figuration of the middle', it is clear from Benjamin's use of tenses in his reading of the photograph that past, present and future are nestled deeply into the image: 'was', 'is' and 'will be' are simultaneously overlaid onto each other (Price, 1994: 40). As Eduardo Cadava has noted, Benjamin suggests elsewhere that the photograph is 'an image in which the Then and the Now come together, in a flash of lightning, in a constellation. In other words, an image is dialectics at a standstill' (Cadava, 1997: 64). In this attempt to understand the temporality of photography, Benjamin is of course not alone; Roland Barthes' reading of the 'anterior future' of the image presents his grappling with the tenses through which to discuss photography's own 'writing with light' (Barthes, 1984: 96).

Indeed, just as Weileder's practice suggests a fascination with the intersection of time and space, so Benjamin suggests that every photograph presents 'space-crossed time'. If the photographic image is 'dialectics at a standstill', then, according to Cadava:

> it interrupts history and opens up another possibility of history, one that spaces time and temporalises space. A force of arrest, the image translates an aspect of time into a certain space, and does so without stopping time, or without preventing time from being time. Within the photograph, time presents itself to us as this 'spacing'. (Cadava, 1997: 61)

Cadava reads Benjamin's concept of 'space-crossed time' as 'time-becoming-space and space-becoming-time' (Cadava, 1997: 61). In a reading of time that is very evocative of Weileder's *Transfer* and *House* projects, and the photographic remembering of a continual process of building and un-building, Cadava suggests that:

> what is spaced here [...] are the always becoming and disappearing moments of time itself. It is precisely this continual process of becoming and disappearing that, for Benjamin, characterises the movement of time. Speaking of Proust, in a passage that asks us to think about the relation between time and space, he writes: 'the eternity that Proust opens to view is space-crossed time, not limitless time. His true interest concerns the passage of time in its most real, that is, its *space-crossed* figure'. (Cadava, 1997: 61, original emphasis)

Proust's writing attempts to represent an experience of time in which the present is occasionally interrupted; and where a space abuts the flow of duration that, in its continuous form, it could not contain. Photographs also present space-crossed time. Cadava describes the 'moment of the photographic event' then as an 'abbreviation that telescopes history into a moment – an abbreviation or miniaturisation that tells us that history can end or break off' (Cadava, 1997: 63). It is this very telescoping that Sugimoto attempts to overcome through his static images of apparently eternal subjects, and that Weileder avoids by expanding the photographic 'abbreviation'.

It is notable that both contemporary artists' experiments with photography's nexus of time and space coincide both with philosophies about the shifting contemporary experience of space-time, and with radical changes in the experience of photography as a medium *per se*. David Harvey's influential analysis has claimed that from the mid nineteenth-century, 'capitalism became embroiled in an incredible phase of massive long-term investment in the conquest of space' (Harvey, 1989: 264). The advent of new technologies, such as the telegraph, radio communication, and of course photography, ran alongside the development of rail travel and steam shipping which sparked a 'subduing of space' and a crisis in the experienced relationship of time and space (Harvey, 1989: 265). Harvey argues that as the new century veered towards the First World War, an annihilation of space through time, increasingly reflected in modernist art and literature such as that of Picasso, Braque, Joyce and Proust, reached crisis point. Just as Benjamin discusses Proust's writing in his concept of 'space-crossed time', so Harvey suggests that Proust's attempt to 'recover past time [...] rested on a conception of experience across a space of time' that was a reaction to the contemporary destruction of space at the hands of increased speed (Harvey, 1989: 267). Harvey diagnoses this capitalist-fuelled modernist transformation in the relationship between time and space as 'time-space compression', and notes that the second, intensive round of it emerged with postmodernism (Harvey, 1989: 283).

This latest period of the 'time-space compression' is identified by Harvey as 'an intense phase' that has had a 'disorienting and disruptive impact upon political-economic practices, the balance of class power, as well as upon cultural and social life' (Harvey, 1989: 284). Harvey claimed that new technologies and organisational forms have accelerated levels of production, exchange and consumption, in turn encouraging the development of globalised mass markets which overcome spatial barriers. The effects of this system, he argued, have been an emphasis upon 'the values and virtues of instantaneity [...] and of disposability' (Harvey, 1989: 286). Harvey's argument is now well known and discussion of its influence has not abated with the decline of postmodernism itself. As the acceleration of globalised capitalism continues, the impact of time-space compression is still declared to be increasingly significant. Paul Virilio has also claimed that our sense of distance has been 'polluted' at the hands of real-time technologies and faster transport and communication devices (Virilio 1997: 58). Most recently in his book *The Futurism of the Instant: Stop-Eject*, Virilio has argued that 'the instant dominates all duration' (Virilio, 2010: 91). The implication is that technological advances – themselves often aimed at projects of endless mapping – alter our experience of both time and space, and in doing so alter our understanding of both past and present:

> let's go back for a moment to the perspective of real time offered by ubiquity, of which Google Earth is just one aspect among others, to this very particular relief that affects not only our subjective and interpersonal relationships, but further, and especially, our connection to the world. With habituation to multiple screens, the focus of the visual field diverts us from peripheral vision, from the open field that gave its everyday fullness to the real space of the verges of our activities and, as a result, causes disorientation in being-there. The teleobjective proximity of transmission tools thereby considerably alters our grasp of the surrounding environment in which each of us physically evolves. (Virilio, 2010: 80–81)

Our horizon line – both temporal and spatial as figured in Weileder's crossed *Seascapes* – is distorted by the ubiquitous screens which allow us to zoom in and out of virtual versions of real-times and spaces. So it is, claims Virilio, that 'historic time' is accelerated to the point at which duration succumbs to the omnipresent instant, thereby enacting an 'assault' on memory:

> this is indeed one of the unacknowledged aspects of globalisation of a real time that subverts not only the real space of the geography of the globe, but also our relationship to time that is really present, since we know from experience: 'it's always questioning the present that causes us to question the past'. (Virilio, 2010: 91–92)

For Virilio, then, 'whether we are dealing with the infinitely big of historicity or the infinitely small of instantaneity', this new lack of experienced duration is a fearful and disorientating condition (Virilio, 2010: 100). Benjamin (1999a: 519) suggested photography captures a 'strange weave of space and time'. Cadava suggests, however, that photography is inextricable from history – 'there can be no thinking of history that is not at the same time a thinking of photography' (Cadava, 1997: xviii). These contrasting views are entirely connected to the acceleration process which Virilio bemoans. Our screens are infiltrated with photographs, our maps constructed out of them, and as they are used in social media sites online, our identities are increasingly shrunk to fit them. Weileder's *Atlas*, in its insistence on adding to the photograph, on splicing time and space together beyond the conventional photographic process, suggests that photography, *per se*, is no longer understood as a 'strange weave of space and time', but is rather a contributing factor to our lack of understanding of physical time or space. Weileder's additions to the photographic process reinforce that crossed

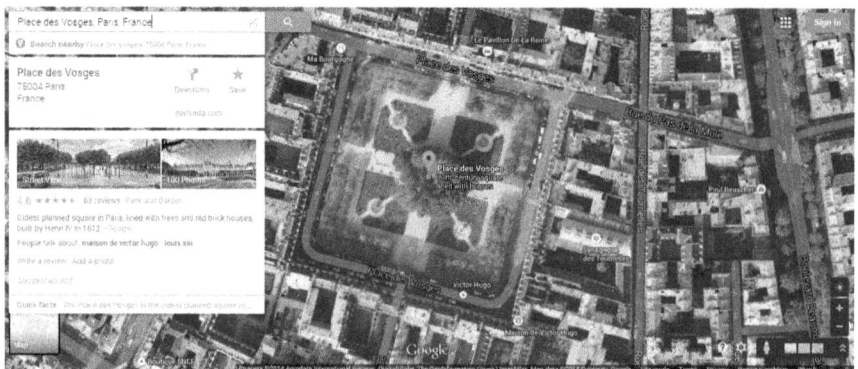

**Figure 5.7** Screen grab from Google Maps, December 2014. Imagery © 2014 Aerodata International Surveys, DigitalGlobe, The Geoinformation Group | InterAtlas, Map data © 2014 Cybercity, Google. This figure has not been made available under a CC licence. Permission to reproduce it must be sought from the copyright holder.

horizon line of space and time that is made visible in the *Seascapes* series: a reassertion of space-crossed time and time-crossed space in an increasingly virtual experience in which those coordinates are rarely mapped over each other. Indeed, even when these coordinates are argued to coexist in digital maps that constantly update according to the user's location, the result is a highly individualised and 'performed' one; Weileder's *Atlas* offers instead a rare shared representation of time overlaid with (public) space which allows sustained and reflective study[2] (see Figures 5.7 and 5.8).

Virilio's argument is certainly pessimistic, and discussion of digital mapping often veers between excitement and concern. In 2009, geographers Paul Kingsbury and John Paul Jones III diagnosed a field which they saw as clearly divided, in a Nietzschian form, between an Apollonian, Adorno-esque fearfulness and dystopia on the one hand, and a Dionysian, hopeful, democratising, participatory vision on the other. Within such a division, Kingsbury and Jones positioned both Benjamin's writing and Google Earth itself, within the latter camp. The nervousness and cynicism which they detect in a conservative approach is contrasted with the much more attractive mix of a Benjaminian giddiness of childhood, love of wandering, and an interest in Surrealism, montage and intoxication (Kingsbury and Jones, 2009: 503). Within this context, Kingsbury and Jones argue that Google Earth actually 'affirms our senses of belonging and our longing to belong' (Kingsbury and Jones, 2009: 510). Certainly contemporary artists have maximised Google Maps' potential to be used in alternative ways to those of traditional map-reading, which often highlight the eruption of idiosyncrasy or even opposition within the flat

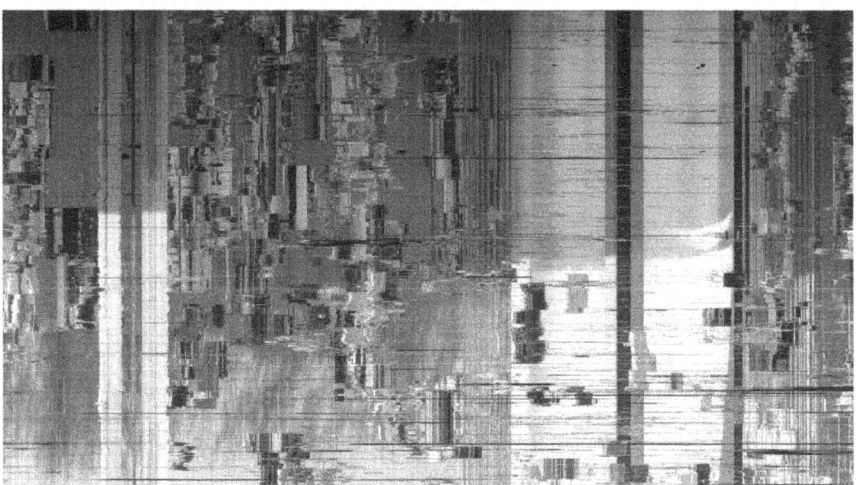

**Figure 5.8** Wolfgang Weileder, *Place des Vosges, Paris, Slice 2356*, 2012, from the *Atlas* series, 2011–present. Archival inkjet print, 137 x 233cm © Wolfgang Weileder (courtesy of Wolfgang Weileder). This figure has not been made available under a CC licence. Permission to reproduce it must be sought from the copyright holder.

recording of the Google camera. Jon Rafman's *The Nine Eyes of Google Street View* (2008–ongoing), for example, famously revealed the artist's search for unusual, beautiful, humorous or illicit moments unwittingly captured by Google's all-seeing eyes (Rafman, 2011; see Figure 5.9). Through the use of screen grabs, his artwork not only isolated and elevated particular Google-generated images from around the world, but in doing so posed questions about which – and how – people, things, places and moments are awarded significance (Rafman, 2011: 7). In this sense, Rafman's project, certainly Dionysian in nature – to use Kingsbury and Jones' analysis – overlaps with some of Weileder's concerns: both artists emphasise the experiential aspect of inhabiting a place at a particular time, and their work serves to disturb or interrupt a distanced, uniform use of 'map as information source'. However, while Rafman's technique of appropriation reveals the persistent visibility of the unexpected or under-represented within Google Street View, Weileder has created his own images and his own *Atlas*, in order to address what he has called 'a dilemma of our time' – how to give ourselves 'anchor points' in a world dominated by scientific and economic flux (Weileder, 2013: 93). This is because for Weileder, 'a map can't reveal how a space is animated by people, weather and light. Our memories and imagined constructions of spaces don't resemble maps' (Weileder, 2013: 93). Even when Rafman highlights the Dionysian animation of Google Street View, his artwork still presents the view of a split-second camera shutter, rather than

**Figure 5.9** Jon Rafman, *D52, Blaru, France*, 2011 from The Nine Eyes of Google Street View © Jon Rafman (courtesy of Jon Rafman). This figure has not been made available under a CC licence. Permission to reproduce it must be sought from the copyright holder.

a built-up and changing memory or experience of a place. Rafman's artwork is reflective of the view of a Google tourist, a Street View surfer; Weileder's is that of a viewer anchored in one space over a period of time, rooted to the ground as slices of time tick past.

Both Weileder and Rafman have created physical art objects that do not deny a Dionysian aspect to online maps and the photography that is used within them. However, each of them also presents deep concerns with these forms of map. Rafman worries about the tendency for Google's cameras to fall upon 'the poor and the marginalised', and sees the artist's role as challenging 'Google's imperial claims' and its 'right to be the only one framing our cognitions and perceptions' (Rafman, 2011: 7). Conversely, Weileder, as I am arguing here, implies that an experience of 'space-crossed time' is absent from such projects. Notably, these critiques do not necessarily place either artist, nor this argument, within the dystopian and fearful category which Kingsbury and Jones outline and dismiss. As a by-product of their argument, Kingsbury and Jones suggest an implicit criticism of interpreting works of art 'as evidence about the societies that made them' – this being a quotation they use from Tim Dean. Instead, they argue for the 'obdurate mystery of cultural and aesthetic artefacts' (Kingsbury and Jones, 2009: 510). In this regard, their snubbing of conservatism seems to break down, as the delights of that which is 'untranslatable to meaning' – another quotation from Dean – take over (Kingsbury and Jones, 2009: 510). This sounds far from Benjaminian, given his famous desire to break free from a ritualistic, cult-valuation of the art object and to instead see it reconfigured within politics (Benjamin, 1999b: 218). While Kingsbury and Jones make a strong case for the Benjaminian, and Dionysian, approach to Google Earth, in the process they risk undermining the value of critique and even of warning, from which Benjamin himself was never afraid to shy away. Weileder's *Atlas* shows no such fear; it is both reflective on the part of the artist and demanding on the part of the viewer in its relationship to contemporary photography and its use within digital maps.

## Time-crossed space

If Weileder's photographic practice emphasises a Benjaminian space-crossed time, his sculptural and installation work presents a similar interest in time-crossed space. Weileder has made many works which map one representation onto another, often (pre)figuring different times and histories across particular spaces. The 2009 installation *Skydeck* consisted of a life-sized reconstruction of the cafe from the next-door Gateshead carpark – a structure that became infamous after its role in the film *Get Carter* – crossing floors and walls in the Workplace gallery after the carpark's demolition. Resting upon scaffolding, and clearly impermanent, the reconstruction was nevertheless clearly identifiable, with the curved corners of the cafe window frames providing a haunting living memory of a former present. The physical space, planned as a home for coffee-drinkers, used as a set for film stars, and recently eradicated, is revived in its new coordinates thanks to Weileder's practice. The Workplace gallery became a space crossed with different times, drawing viewers' attention to the physicality of their memories and the temporality of their present.

Other projects, such as *Le Terme* in Milan, 2008, also present a resurrected former building in a public space (see Figure 5.11). *Le Terme* saw the continual construction and deconstruction of a full-scale replica of the Diurno Venezea, a 1930s public bath-house that was situated underneath the Piazza Oberdan. The process was captured on photographic film with one very long exposure. As with *Skydeck*, the ghostly emergence of a former present is striking, in part, because of its crossing of time with space: an area which had seen the bathing of hundreds of now long-departed people is made to co-exist in the present with current Milanese inhabitants and shoppers: times coalesce in the same space. There are resonances with other ambitious artistic projects concerned with

**Figure 5.10** Wolfgang Weileder, *Skydeck*, 2009, installation in the Workplace Gallery, Gateshead © Wolfgang Weileder (courtesy of Wolfgang Weileder). This figure has not been made available under a CC licence. Permission to reproduce it must be sought from the copyright holder.

memory and lived space: Rachel Whiteread's cast objects, rooms and even an entire *House* (1993) also offer a poignancy of past merging with present, and Elizabeth Wright's *Installation, Bungalow Showroom Gallery* (1996), an immaculate replica of a planned but never realised home, presents a similar life-sized rendering of different timescales, realising a future that was intended but never built. Weileder's installations provide a sculptural, spatial rendering of his photographic processes: a desire to infiltrate lived space and time with each other, and to present an experience that is neither the 'infinitely big of historicity nor the infinitely small of instantaneity' (Wells, 2013b: 105), but rather the life-size nature of duration.

Indeed, the scale of Weileder's projects demonstrates a consistent preoccupation with the life-size. The 2012 *Res Publica* continued the artist's concern with public space and architecture while also investigating the relationship between space and time. Weileder first designed and made a silver leaflet-stand, mirroring the Palladian architecture of Washington's Supreme Court of Justice (see

'Space-crossed time'  131

**Figure 5.11** Wolfgang Weileder, *La Terme, Piazza Oberdan, Milan*, 2008 © Wolfgang Weileder (courtesy of Wolfgang Weileder). This figure has not been made available under a CC licence. Permission to reproduce it must be sought from the copyright holder.

Figure 5.12). The leaflet-stand, similar in concept to those offering information on real estate, offered passers-by a free sheet containing the plans for a cardboard scale model of the Supreme Court. In effect, Weileder's leaflet-stand, positioned on a street corner by Cass Gilbert's imposing Palladian building, offered people in Washington the opportunity to create their own cardboard Supreme Court at a scale of 1:50. Local art students created some models from the plans, which were then positioned at key points throughout the city and left to the elements. Over time, the human scaled-models became battered, crumpled and damp. Some were removed, others left slumped around the city, their cardboard Beaux-Arts pillars buckling under the effects of both time and bad weather (see Figure 5.13).

In its conjunction of two very different architectures – the symbolic power of a vast building constructed to house and distribute justice, and the human-scaled, fragile and vulnerable cover for a sheltering body – *Res Publica* posed pointed questions about the nature of justice and freedom in a Western society in which many live without a home (Wells, 2012: 77–89). That it did so through miniaturising the Supreme Court to a human-scaled model, is key. The Supreme Court, vast as it is, is primarily symbolic, a façade demonstrating corrective

**Figure 5.12** Wolfgang Weileder, *Res Publica*, 2012, 5x5 Public Art Festival, Washington DC © Wolfgang Weileder (courtesy of Wolfgang Weileder). This figure has not been made available under a CC licence. Permission to reproduce it must be sought from the copyright holder.

power and supreme authority. The bodily-sized cardboard version is much closer to lived experience, marking the spatial limits of our physical existence, as well as our temporal vulnerability to decay.³ While the architecture of the Supreme Court looks back to a historic Greek past and claims its indelible vast placement in the present, in the model, real-time and real-scale are overlaid. To use Virilio's terms, here again, Weileder draws attention away both from the mighty grandeur and the 'infinitely big of historicity', and from the weightless flippancy and the 'infinitely small of instantaneity' (Wells, 2013b: 105). The focus is again on a time-crossed space, and an embodied experience of duration that, in the case of homelessness as the crumpled cardboard eloquently suggests, is not represented justly either in the grand permanence of our national historic symbols or in the flashes of representation that flicker within a feed of fast-flowing information.

That Weileder's work spans both photography and sculpture is perhaps a natural accompaniment for this sustained interest in the relationship between space and time. His 2002 project in Kielder Water and Forest Park, UK, brings both aspects of his practice together; creating a sculpture that could photograph. *Camera* (2002) consisted of two blue large-scale tents, one turned on

**Figure 5.13** Wolfgang Weileder, *Res Publica*, 2012, 5x5 Public Art Festival, Washington DC © Wolfgang Weileder (courtesy of Wolfgang Weileder). This figure has not been made available under a CC licence. Permission to reproduce it must be sought from the copyright holder.

its head and intersected with the other (see Figures 5.14 and 5.15). While the grounded tent functioned as a viewing space, the upturned one was transformed into a camera obscura, with mirrors positioned inside so as to turn the resulting image at the same angle as that of the tent. Given the artificial nature of the Northumberland park, Weileder's manipulation of the natural 'trick' of nature to 'reproduce itself', as early pioneers described photography, resonates with the long-discussed 'naturalness' or artificiality of photography as a medium.[4] The effect for viewers entering the first tent would be to imagine a shift in their own ground, as the environment around them seemed tilted to the jaunty angle of the second tent. Accompanying this disorientation, though, is an overlapping, a mapping onto each other, of significant coordinates in time and space, and in sculpture and photography. Here the human shelter, the bodily tent, is at one and the same time the camera, the room of the image. Three-dimensional, liveable space is itself the producer of the image on the screen, an image that reflects the real-time duration unfolding in the physical environment beyond. Duration and location are united: space crosses time and time crosses space.

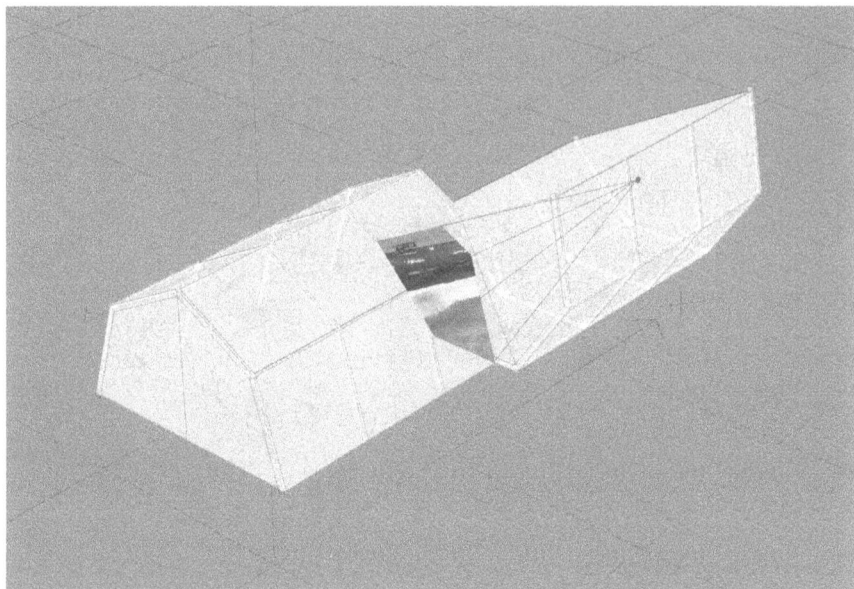

**Figure 5.14** Wolfgang Weileder, *Camera*, 2002 © Wolfgang Weileder (courtesy of Wolfgang Weileder). This figure has not been made available under a CC licence. Permission to reproduce it must be sought from the copyright holder.

**Figure 5.15** Wolfgang Weileder, *Camera*, 2002. Fabric, aluminium, optics, 220 x 780 x 420cm © Wolfgang Weileder (courtesy of Wolfgang Weileder). This figure has not been made available under a CC licence. Permission to reproduce it must be sought from the copyright holder.

Weileder's work, then, from site-specific installation to digital photographic composite, resists both the infinitely big and the infinitely small, just as it resists the fixity of historicism and the fleeting nature of the instant. As such, it suggests that the conventional time-space coordinates of both photography and public space are no longer adequate in relaying, or reminding us, of the nature of 'space-crossed time'. In the context of a society increasingly distracted by the omnipresent instant of the image, Weileder's work suggests that the former auratic weave of space and time within both photographs and memories of places is miniaturised, reducing time and space to flattened, static images. Instead, Weileder's work demonstrates the need to construct, to extend, to build upon, current experience, thereby offering a renewed experience of space-crossed time.

In this respect, Weileder's practice echoes another of Benjamin's key ideas about photography – that 'constructive' rather than 'creative' photography is necessary in order to 'reveal anything about reality' (Benjamin, 1999a: 526). While the 'creative' detaches itself from 'physiognomic, political and scientific interest', and ends up merely 'capitulating to fashion' or advertising, 'constructive' photography recognises that 'something must be built up, something artificial, posed' in order to communicate meaning (Benjamin, 1999a: 526). For Benjamin, of course, this meaning is often that of the historical materialist, who searches for the reification of human relations to be made explicit. Weileder's photography, and indeed his sculpture, suggest both such a constructed, built nature, and also an acknowledgement of this aspect of their condition. For Benjamin, 'constructive' photography, that which is given to 'experiment and instruct' rather than to 'charm or persuade', is of most value (Benjamin, 1999a: 526). It leads him to close his 'Little history of photography' essay with László Moholy-Nagy's foresight that 'the illiteracy of the future will be ignorance not of reading or writing, but of photography' (Benjamin, 1999a: 527). Weileder's constructive photography and installation suggest that the contemporary capitalist culture of the instant image is producing a form of illiteracy in terms of experiencing and understanding the nexus between time and space. Further, his work remembers and reminds viewers of an existence in the here and now of public space which, as Virilio argues, can affect our social and ontological activity as much as our individual experience (Virilio, 2010: 80). Weileder's artwork to date then, with its overlapping cross of time and space, pinpoints the intersection of the real-time and the life-size. In the *atlas-project* and throughout his oeuvre, Weileder constructs a contemporary conception of space-crossed time that is determinedly human-scaled, and which, in unmasking the distractions of instantaneous creative images, offers unsentimental reference points for locating our own spatio-temporal condition.

## Notes

1  Aspects of this chapter were published in Wells (2013b).
2  Nanna Verhoeff has suggested that mobile digital cartography enables a shift from the representation to the performance of space. She notes that in the process, the potential pitfalls of representation are avoided, and the viewer is repositioned as central (Bounegru, 2009).
3  Weileder's interest in working to a 1:1 scale places him within a contemporary trend of artwork exploring the effects of scale (Wells, 2013a).
4  In 1839, Louis Daguerre famously described his invention of the Daguerreotype as a means by which 'nature' could 'reproduce herself'.

## References

Barthes, R. (1984) *Camera Lucida*. Translated by R. Howard. London: Fontana.
Benjamin, W. (1999a) 'Little history of photography'. In: Jennings, M. W., Eiland, H. and Smith, G. (eds) *Benjamin: Selected Writings, Vol. 2, Part 2, 1931–1934*. London: Belknap Harvard, pp. 507–531.
Benjamin, W. (1999b) 'The work of art in the age of mechanical reproduction'. In: Arendt, H. (ed.) *Illuminations: Walter Benjamin*. London: Pimlico, pp. 211–245.
Bounegru, L. (2009) *Nanna Verhoeff: Mobile Digital Cartography from Representation to Performance of Space*. [Online] Available at: http://networkcultures.org/urbanscreens/2009/12/05/nanna-verhoeff-mobile-digital-cartography-from-representation-to-performance-of-space/ (accessed 1 March 2015).
Cadava, E. (1997) *Writing with Light: Theses on the Photography of History*. Princeton, New Jersey: Princeton University Press.
Cosgrove, D. (2008) *Geography and Vision: Seeing, Imagining and Representing the World*. London: I. B. Tauris.
Harvey, D. (1989) *The Condition of Postmodernity: An Enquiry into the Origins of Cultural Change*. Oxford: Blackwell.
Kingsbury, P. and Jones III, J. P. (2009) Walter Benjamin's Dionysian adventures on Google Earth. *Geoforum*, 40: pp. 502–513.
Price, M. (1994) *The Photograph: A Strange, Confined Space*. Stanford, California: Stanford University Press.
Proust, M. (2002) *In Search of Lost Time, Volume 1*. Translated by C. K. S. Moncrieff, London: Vintage.
Rafman, J. (2011) *The Nine Eyes of Google Street View*. Paris: Jean Boîte Éditions.
Sontag, S. (1979) *On Photography*. London: Penguin.
Sugimoto, H. (2010) *Hiroshi Sugimoto*. Berlin: Hatje Cantz Verlag.
Virilio, P. (1997) *Open Sky*. Translated by J. Rose. London and New York: Verso.
Virilio, P. (2010) *The Futurism of the Instant: Stop-Eject*. Translated by J. Rose. Cambridge: Polity Press.

Weileder, W. (2013) 'An interview between Wolfgang Weileder and Alistair Robinson'. In: Seyfarth, L., Wells, R. and Robinson, A. (eds) *Wolfgang Weileder: Continuum*. New York: Kerber Verlag.

Wells, R. (2012) 'Housing justice: Wolfgang Weileder's *Res Publica*'. In: Hollinshead, R. (ed.) *Magnificent Distance*. Newcastle: Grit and Pearl, pp. 74–89.

Wells, R. (2013a) *Scale in Contemporary Sculpture: Enlargement, Miniaturisation and the Life-size*. Farnham: Ashgate Publishing.

Wells, R. (2013b) 'Space-crossed time': Mapping the co-ordinates of Wolfgang Weileder's photography and installation'. In: Seyfarth, L., Wells, R. and Robinson, A. (eds) *Wolfgang Weileder: Continuum*. New York: Kerber Verlag, pp. 105–141.

# 6

# Traces, tiles and fleeting moments: art and the temporalities of geomedia

*Gavin MacDonald*

## Introduction: geomediation in the inhabitable map

In this chapter, I discuss ways in which artists have exploited and exposed the temporalities of 'geomedia'. I am following writers working at the intersection of media studies and geography in using this term to refer to a contemporary complex of technologies, content and practices that involve mapping, remote survey visualisations and the binding of digital information to location via GPS (Thielmann, 2010; Lapenta, 2011). My aim is to challenge the idea that geomedia's only temporal effects are ones of timelessness. Crucial to this challenge is a separation out of two different, and sometimes conflated, versions of timelessness ascribed to cartography and new media. However, rather than an artwork, I will begin with an image of artists *at* work. Depicting a moorland landscape split by a narrow, unmarked road, and with two figures in the midground to the right, this panoramic digital photograph is one image among the millions of others that, since 2007, have been captured and stitched into the dynamic archive of photographic mapping that is Google Street View.[1] Over the course of their lives, most of these automated images – instances of what Joanna Zylinska calls 'non-human photography' – will escape serious human scrutiny (Zylinska, 2013). Until April 2014, these lives would have been cut short by updates from the restless fleet of Google's Street View cars. Since the introduction of a time slider to Street View's interface, however, such images remain accessible; although with all the millions of miles of road covered, and given the evident remoteness of the location depicted, it is unlikely that I would have found this particular image if I had not been told precisely where to look. This image crystallises something important about several matters at stake here: geomedia's ubiquity, the way it has become part of our 'technological unconscious', to use Nigel Thrift's term for the performative infrastructures which pattern our everyday lives (Thrift, 2004: 175); the remarkably short timescale these developments have unfolded over; the place of art in addressing and exploring these changes; and lastly the relationship of the photograph to the map. As I set out in this chapter, issues of temporality are key to that relationship, both practically – in the photographic mapping services that Google provides – and analogically.

The figures captured in this photograph are those of the Dutch media artist Esther Polak and her partner Ivar van Bekkum; it was taken in September 2009 during their residency at the Highland Institute for Contemporary Art, an artist-run space in a converted farm building to the south of Loch Ness. When the Street View car passed them, Polak and van Bekkum were releasing GPS-equipped balloons with the intention of visualising their journeys in Google Earth (personal correspondence, 11 January 2011). Polak is one of the best-known artists to use GPS and to work with the motif of the mapped trace of movement. Her 2002 work, *Amsterdam RealTime*, re-imagined GPS as a tool for a bottom-up, participatory cartography: over two months, beginning with an initially black projection screen, a map of that city was built up through the movements of sixty of its inhabitants (Figure 6.1).[2] *Amsterdam RealTime* is a key early work in the 'locative media' genre: this label was first used in the early to mid-2000s by a community of practice working with GPS in art and open data activism (Hemment, 2006; Tuters and Varnelis, 2006; Zeffiro, 2012; O'Rourke, 2013; Pinder, 2013). This label is still used in this sense and in academic literatures as a synonym for augmented reality, geomedia and neogeography (Crampton, 2009; Thielmann, 2010; Lapenta, 2011). Though GPS had been sporadically used by artists during the 1990s, the US government's suspension of the purposeful degradation of the consumer signal in May 2000, a function known as 'Selective Availability', meant accurate GPS receivers became cheaply available.[3] This was a crucial move in establishing GPS as a ubiquitous consumer technology, and it opened a space for creative activity.

For Polak and the other figures associated with 'locative media' as it emerged in the early 2000s, technologies had to be hacked together: *Amsterdam RealTime*'s participants carried a satchel with a PDA (Personal Digital Assistant) phone attached to a GPS signal receiver that required an externally worn signal-boosting aerial, and visualising software had to be purposefully written for the project. As I write more than a decade later, most of its functionality could be achieved with off-the-shelf smartphones, apps and web mapping services. The fact that by 2009, Polak was using Google Earth in her projects to visualise traces of movement, speaks of the thorough mainstreaming of GPS. Polak and

**Figure 6.1** *Amsterdam RealTime* (2002) Esther Polak, Jeroen Kee and Waag Society. Participatory mapping project and installation; this image depicts a composite of GPS traces produced by participants (courtesy of Esther Polak). This figure has not been made available under a CC licence. Permission to reproduce it must be sought from the copyright holder.

van Bekkum liken their encounter with the GPS-equipped Street View car to the 'Droste effect': named after a famous poster advertising a brand of cocoa, this is a Dutch term for *mise-en-abyme*, the infinite recursion of an image that contains a reproduction of itself (personal correspondence, 11 January 2011).

Geomedia is a category that encompasses the cartographic visualisations of moving bodies, produced by: Polak in *Amsterdam RealTime*, the Street View photography that captured her at work in Inverness-shire seven years later, and the vast and complex techno-social assemblage that has made both those visualisations possible. A brief list of some of the most significant elements of geomedia would certainly include the following: GPS and mobile computing; the georeferencing of digital photography and other traditional media formats; the ubiquity of maps as interfaces for data; and the widespread access we have to ceaselessly (if unevenly) renewing archives of survey photography – photography which assumes both the Gods' eye view of satellites and the view at ground level. If the media art that was gathered around the 'locative media' label in the early to mid-2000s, explored and sought critical and alternative versions of a technological imaginary that was still coming into being,[4] the 'Droste effect' Street View capture of Polak and van Bekkum in 2009 and their embrace of off-the-shelf technology for their practice, speak of a world in which such technologies and visualisations have been fully woven into our everyday lives, commercialised, banalised even. Seven years from Polak's initial foray into mapping as art, four years from the launch of Google Maps and Google Earth, and two since the launch of Street View, in 2009 we were already living in a world where digital maps and technologies of tracing had receded from bleeding-edge novelty into the technological unconscious. Thrift's name for this world is 'the inhabitable map' (Thrift, 2011: 9).

This purported inhabitability is a problem: as Valerie November and her co-authors argue in their unpacking of the new understandings of space produced by digital mapping technologies and practices, '*no one* and *no thing* ever resided in the virtual image of the map' (November, Camacho-Hübner, and Latour, 2010: 594). The map, as it has been imagined in the Western scientific tradition, is lifeless; static; denying processes and temporality; cleaving time from space to construct it. As Doreen Massey put it, the map is 'the sphere of a completed horizontality' (Massey, 2005: 107). Denis Wood and John Fels take a similar position when they say that 'most maps exist in the present, or if they can possibly get away with it, the aorist: no duration at all ... out of chronology ... free of time': the 'aorist' is the Greek simple past tense, referring to completed actions with no indication of their current status or their durative thickness (Wood and Fels, 1992: 112). Wood and Massey are in agreement that the map represents an already completed present that is mute about the unfolding of events in the past, and that forecloses the future. As Massey memorably puts it, 'loose ends and ongoing stories are real challenges to cartography' (Massey, 2005: 107).

## Two kinds of timelessness

To complicate matters, some writers on geomedia are concerned that there is not one kind of atemporality at stake, but two; the closure of the map to past and future is paralleled by, and conflated with, a timelessness associated with networked information and communication technologies (Farman, 2012; Graham, Zook, and Boulton, 2013). This second concept of atemporality can be found in prototype in Henri Bergson's and Martin Heidegger's respective critiques of 'spatialised' or 'inauthentic' time (which is to say, clock time) (Mackenzie, 2002; Mitchell and Hansen, 2010). But, if under modernity the technical mediation of time and its standardisation was already seen as alienating humanity from the authentic experience of time, writers on technology and society in the closing years of the twentieth century described a dramatic acceleration of this effect. For critics like Paul Virilio, the instantaneity of the real-time communication systems that grant us global awareness and action at a distance, belies computational time's destructive effect on subjective, experiential time (Virilio, 1997; Mitchell and Hansen, 2010). Speaking broadly about this attitude, Adrian Mackenzie describes how 'technological speed can give the impression that the future is closed, and that any experience of time grounded in duration and memory has been lost' (2002: 1). For Manuel Castells, writing in the mid-1990s with then-current critiques of the postmodern in mind (particularly David Harvey's (1989) *The Condition of Postmodernity*), 'the culture of real virtuality associated with an electronically integrated multimedia system' produces timelessness, both through the temporal immediacy of distant events and information and through the 'mixing of times in the media'. Producers and users have access to an undifferentiated history of cultural expressions and can combine these in temporal collages 'where not only genres are mixed, but their timing becomes synchronous in a flat horizon, with no beginning, no end, no sequence' (Castells, 1996: 461–462). 'Timelessness', Castells tells us, 'is the recurrent theme of our age's cultural expressions' (Castells, 1996: 463).

## Time, space and the mapped trace

These ideas about timelessness depend on an opposition between technology and society. In his study *Transductions*, Adrian Mackenzie (2002) shows how that opposition has been challenged by relational approaches to technology, where to be human is to be necessarily bound up with technology, as with Bernard Stiegler's employment of Jacque Derrida's concept of 'originary technicity'.[5] For Stiegler (1998), the tools with which we mediate time and space specify the way in which we experience them. Mackenzie (2002) describes how related ideas – though more focused on the 'lives' of technologies themselves – were developed earlier by the philosopher Gilbert Simondon; for Simondon, technologies themselves form something like the perimeter of human societies, the point at which properties from physical contexts (properties of matter which produce particular effects, things like hardness or conductivity) are enfolded into our collectives and made portable and repeatable (Simondon, 1980). Simondon uses technicity to describe the degree to which the useful, repeatable effects of technical objects are concretised within networks of relations; these effects are extended both spatially (in reproduced objects) and temporally in changing, historically specific manifestations. Technicity therefore also 'refers to the historical mode of existence of technical mediations': objectivations of particular lineages of technicity change over time in terms of the make-up of the milieus whose effects they enfold (Mackenzie, 2002: 108).

These ideas can be illustrated through Mackenzie's discussion of GPS, and his alignment of it with an earlier objectivation of the technicity of clocktime; the pendulum clock invented by the seventeenth-century Dutch polymath Christiaan Huygens. For Mackenzie, both pendulum clock and GPS confirm 'the inseparability of timing and spacing': to produce positional fixes and measure movement through space, the constellation of GPS satellites bathes the world in a broadcast of Coordinated Universal Time (Mackenzie, 2002: 108). Huygens proposed that the fixed length of his pendulum be used as the basis of a standard global unit of measurement, the 'hour-foot' (Huygens, 1986, cited in Mackenzie, 2002: 101). Apart from the matter of the intervening three centuries, Mackenzie (2002) argues that what separates these two technical objects is a difference in their respective definitions of a second, the minimal unit of time. Whereas Huygens' second was based on the period that a pendulum of a defined length took to complete a movement, the clock signals of GPS are based on the wavelength of the microwaves emitted by the caesium atoms in the satellite's onboard atomic clocks (9,192,631,700Hz). Mackenzie notes that the move from one 'oscillation' per second to more than nine billion could be interpreted, as per Virilio (1997) and Castells (1996), as evidence of the inexorable acceleration of technology (seen as a monolithic abstract entity) beyond the limits of human sense-experience, 'a symptom of how human time is being lost to an inhuman, globalising, technological "time"' (Mackenzie, 2002: 88). However, technicity in Simondon's sense provides a way of thinking about this acceleration without resorting to pessimistic narratives based on an abstract opposition of society and technology. The regular rate of repetition known as a second derives in both cases from what new media theorist Mark B. N. Hansen calls 'a distinct absorption of material contingency' (Massey, 2005: 1210), in the ensembles within which pendulum clocks and GPS, as historically specific concretisations of the technicity of time, are reproduced: the planet's gravity for the former and the electromagnetic fields emitted by caesium atoms for the latter.

Hansen argues that mobile GPS devices are producing a practical understanding of 'the interrelation of time and space, one in which time and space lose their heterogeneity as they become more and more intensely imbricated with one another', and similar positions can be found elsewhere in the literature on digital mapping and 'locative media' (Hansen, 2005: 1207; November, Camacho-Hübner, and Latour, 2010; Speed, 2011). From this, it seems like GPS has the potential to support Doreen Massey's project to reimagine space and time as interpenetrating rather than counterposed, the sphere of a plurality of interweaving trajectories – a dynamic simultaneity rather than a completed horizontality (Massey, 2005). If – as in Michel de Certeau's account – the early modern map removed narratives of travel as it worked towards the status of a science, becoming 'a totalising stage on which elements of diverse origin are brought together to form the tableau of a "state" of geographical knowledge', then the GPS trace can be seen as a return to the map of 'the operations of which it is the result or the necessary condition' (de Certeau, 1984: 121).

Writing with W. J .T. Mitchell, Hansen illustrates the way that GPS brings about 'a concrete suturing of time and space' through a discussion of *Landing Home in Geneva* (2005), a GPS-based artwork by Masaki Fujihata from that artist's long-running *Field Works* series (Mitchell and Hansen, 2010).[6] This work involved interviews with immigrants to that city, conducted while walking from their homes to a place which felt like home to them: these walks were filmed using a panoramic lens and the subsequent footage was animated as a cylindrical form, tracking along a line denoting the trace of the walk, all modelled in a virtual, three-dimensional space. Mitchell and Hansen claim that:

> Fujihata shows us how, in the midst of a rapidly accelerating surveillance society, we can use the new found technical precision of space-time mapping as a rich and poignant means of asserting our own existential uniqueness. His media-specific configuration of time, space and embodiment gives us the opportunity to map global space-time in relation to our own movement through it. (Mitchell and Hansen, 2010: 110)

As well as issues of power and control, artworks that employ these technologies enable us to reflect on our own participation in acts of timing and spacing,[7] our own roles as moving parts in a socio-technological ensemble of satellites tracking in their orbits, radio signals and packets of data, among other things.

## Time and photographic mapping

In their analysis of the temporal codes of mapping, Denis Wood and John Fels make a parallel with the snapshot photograph, likening what they perceive as an atemporality in snapshots to the way in which we understand the map: 'in time but not of it; something with the time evaporated out of it' (Wood and Fels, 1992: 126). This analogy is inappropriate; snapshot photographs may record punctual events extracted out of meaningful sequence, a momentary registration of light on a sensitive surface, but this doesn't amount to timelessness. Their use as supports for memory means that temporality is rarely absent from the ways in which we interpret and are affected by them. In any case, Wood and Fels go on to soften their stance, saying, 'we *may* be aware of emplacing time in the photograph … but we refuse to extend these understandings to the map. Time remains a … hidden dimension' (Wood and Fels, 1992: 126).

Wood and Fels formulated their semiotic framework for understanding maps during the 1980s, and subsequently incorporated it into *The Power of Maps* (Wood and Fels, 1992). Wood's critical treaties on cartography were accompanied by a discussion of a particular image that, for Wood, seemed to aspire to the condition of cartography and also served as a snapshot at the same time. Stressing the importance of temporal selections and combinations, Wood narrates the production of Tom Van Sant's composite satellite image of the world from space; an image used as the half title page illustration for the 1990 *National Geographic Atlas of the World* and subsequently in 1992 as a poster to promote the environmental efforts of a telecommunications company, under the title 'A Clear Day'. This is a world where it is always daylight, always cloudless, always high summer: these selections are key to Wood's argument that Sant's image is better thought of as a map than a photograph. Wood goes on to relate the rhetoric of this image to the environmentalist discourse that framed the reception of the photographs of the Earth from space taken by the Project Apollo astronauts.[8]

Wood's analysis works for Van Sant's image, but returning to the present, it needs to be updated. Martin Dodge and Chris Perkins – drawing on much the same temporal markers as Wood – identify an illusory effect of seamlessness and transparency in composite satellite imagery, however, this effect is far from guaranteed (Dodge and Perkins, 2009). While the images of the Earth they provide might be coherent at some scales and framings, applications and services like Google Earth and Google Maps do not hide their temporality, as such – even cursory play reveals the asynchronicity of image tiles, while the interfaces of both Google Earth and Street View explicitly provide the date the imagery was taken and supply time sliders to explore its thickness. Discussing Google Earth, Mark Dorrian (2013) also notes the consistency of cloudlessness and eternal daylight, but argues that despite these unifying criteria, it is still an evidently fragmented, constructed image. For Dorrian the transparency and unity implied by those criteria speak of Google Earth's coherence as an interface for data – 'the "wholeness of its searchability"' rather than its wholeness as a 'natural' image' of an object (Dorrian, 2013: 299). Unlike Wood, Dorrian sees a significant difference between the cloud-swirled blue marbles of the Apollo photographs and the cloudless, eternal day of mosaicked satellite imagery; he argues that these patchwork worlds are a new kind of political map where differing resolutions and image upload frequencies speak more of Western political, security and economic interests, wherever they may lie, than they do of the bounded territories of nation-states.

Regardless of the overall effect of temporal patchworking in satellite imagery, time is emplaced in the conventional way within the unique images that comprise it. Artists have exploited these archives of fleeting moments in ways that draw their affective potential from the tension between the historical specificity of the event, or conditions depicted, and the idea of continual, exhaustive surveillance and recording represented by geomedia. The artist Laura Kurgan has spent the last two decades critically exploring the cultural implications of different digital mapping and remote imaging technologies; in 1994 she was responsible for one of the first artworks to use GPS, the installation *You Are Here: Information Drift*.[9] For *Monochrome Landscapes* (2004) Kurgan purchased commercial Earth observation satellite imagery of four eight-by-eight kilometre locations dominated by one of the colours white, blue, green and yellow, and the results were exhibited as large Cibachrome prints with the pixellation of the satellite's scanner array clearly visible.[10] As Kurgan relates, she 'was interested in the idea that the places on earth that appeared from above as more or less a single colour were also places that were contested, fragile, and subjected to an increasingly thorough surveillance apparatus' (Kurgan, 2013: 153). All the locations were environmentally or politically sensitive, and for the colour yellow a chance encounter was captured that helps to locate the image in geopolitical time: crawling bug-like over the monochrome expanse of a section of the Iraqi desert, two weeks into Operation Iraqi Freedom, two military helicopters can be seen (Figure 6.2).[11]

The artist Mishka Henner has captured a more varied and visceral palette of colours in a recent series of images culled from Google Earth, which deals with landscapes that have been shaped by the production of meat. In *Feedlots* (2013), Henner focuses on the US beef industry and the sites where young cattle are fattened before slaughter – an intensely managed and scientifically accelerated process that, alongside a programme of food and supplements, involves a tight spatial organisation of pens, run-off channels and lagoons where the animals' waste products are broken down by chemicals.[12] Order and ordure are counterposed through the proximity of repeated rectilinear forms and roiling abstraction. In a widely reproduced image of the Coronado Feeders lot near Dalhart, Texas, the waste lagoon looks like an orifice or wound, wet and red with gangrenous greens, yellows and blacks crusting around its puckered edges (Figure 6.3). Time is emplaced in this image in a fundamental way, one that

Traces, tiles and fleeting moments 147

**Figure 6.2** Detail from *Monochrome Landscapes* (2004) Laura Kurgan. Yellow: southern desert, south-eastern Iraq, between Al Busayyah and An Nasiriyah. Image courtesy of Laura Kurgan. Includes material © 2003 DigitalGlobe, all rights reserved. This figure has not been made available under a CC licence. Permission to reproduce it must be sought from the copyright holder.

**Figure 6.3** Coronado Feeders, Dalhart, Texas, from *Feedlots* (2013) Mishka Henner. Google Maps.

becomes apparent when exploring the Coronado site using Google Earth's time slider: the image was taken in September 2011, and the lagoon's colours and the contrast with the dun surface of the pens are due to the devastating drought that Texas suffered that year, the worst in over a century (Mishka Henner, personal correspondence with the author, 8 April, 2014). With the water evaporated, the organic and chemical contents in the lagoons were concentrated to lurid effect.

Coming back down to Earth again, it is extremely difficult to see the panoramas of Street View as timeless, strung out punctually as they are along the traces of actual journeys. Like *Amsterdam RealTime* and *Landing Home in Geneva*, they too write movement, and thus temporality, back into the map that has for so long denied them. Street View doesn't just register the more-or-less static form of our built and natural environments, it is an archive of ephemeral events, and several artists have drawn on it in curatorial selections and framings (Moakley, 2012). Jon Rafman's *9-Eyes* project (2009–ongoing), for example, brings together curious, sinister or poignant scenes alongside the uncanny effects produced by glitches.[13] Mishka Henner has also worked with Street View, but with tighter criteria than Rafman. For his *No Man's Land* books, Henner took locations gleaned from discussion forums used by the clients of sex workers and used Street View to source images of them, mainly urban edgeland sites in southern Europe (Henner, 2011; 2012). Many of these images show the moment when the sex workers – all women – look up to address the photographic apparatus, their faces pixelated by Google's pattern recognition algorithm. However, the most poignant of the *No Man's Land* images capture these scenes at those times when they are empty of human subjects: Henner bookends the first volume with images that show only chairs by the roadside.

In a recent analysis of 'the power relations involved in the practices that enact augmented realities', Mark Graham and his co-authors Matthew Zook and Andrew Boulton have discussed Street View photographic mapping and the sharing of georeferenced photographs on Google Maps, among other data overlays (Graham, Zook, and Boulton, 2013: 465).[14] Developing a heuristic typology of those relations, the authors identify temporal power as one of two dimensions which usually pass unnoticed in augmented realities; they place it alongside the power of code, particularly the search algorithms that regulate conduct by regulating the visibility of content. Graham, Zook and Boulton describe 'timeless power' as a 'flattening of time' (Graham, Zook, and Boulton, 2013: 470), but rather than rehearsing the cartographic version of timelessness, they instead refer to Castell's cultural and technological concept of timeless time; likening the excessive, multiple temporalities of photographic mapping to the disordering of chronology that Castells sees as characteristic of contemporary culture. Like Dodge and Perkins (2009), Graham and his co-authors also invoke the idea of seamlessness, this time in relation to representations of place; this wholeness is only made possible by the way in which augmented reality interfaces ignore temporal data while stitching together their overlays. Despite this, one of their case studies is a distinctly seam-like glitch in Google Street View: a street in Lexington, Kentucky, where travelling down Main Street in opposite directions reveals the before and after of a contested example of urban regeneration, a development known as CentrePointe. What is more, on Google Maps the CentrePointe site is peppered by user-contributed photographs that commemorate a demolished music hall. These before-and-after juxtapositions and the pinning, by users, of memories to the map are read by Graham, Zook and Boulton (2013) as a collapse of temporality.

The idea that a surfeit of temporal markers necessarily leads to such a collapse is challenged by a recent artwork that deliberately exploits the temporal seams of Google Street View to reflect on the clearance and gentrification of cities in much the same way as the Lexington Main Street seam. Justin Blinder's *Vacated* (2014) exploits a quirk of Street View's update frequency to capture change in New York's built environment, namely the fact that major roads are updated more regularly than minor ones.[15] The artist used the New York City (NYC) Department of City Planning's PLUTO dataset to find corner lots where new buildings had been added in the previous four years, and then used the panoramas taken at those intersections to make a series of still and moving image works, including a series of animated GIFs which travel years in time as they shuffle a few metres in space.

Francesco Lapenta (2011) offers a less dramatic account of the effect of combining of temporal markers in geomedia that bring together photographic 'elements of diverse origin' (de Certeau, 1984: 121) in static tableaux:

> this synthesised image, as the single photographs that compose it, is still a visible token of the past; but as a combination of contiguous photographs, it is also a combination of past, present and future, as the photographs that it merges together represent the present, past or future to one another. This new digital photographic map transforms a time–space unicum (the photograph taken at a specific time, in a specific place) into a fractured time within a space continuum (a composed photographic image that merges different times and connects contiguous spaces). (Lapenta, 2011: 17)

From the point of view of photography alone, this isn't particularly novel: yes, the imagery these services provide involves abandoning the traditional understanding of the relationship of photography to a unique spatio-temporal referent, but then so does any photocollage technique. What is most at stake here concerns the relationship, practical and analogical, between the photograph and the map, and in particular the challenge that the excessive, visibly fractured temporality of photographic mapping poses to the lingering notion of the map as having a snapshot (a)temporality, in Wood's terms, or of being 'a slice through time', in Massey's words (2005: 107). The stress on atemporality effects in studies of the aesthetics and cultural impact of geomedia, and the conflation of the cartographic and technological/cultural versions of timelessness in particular, moves too quickly to read a surfeit of temporal markers as a collapse of temporality as such. In much the same way that the mapped trace of moving bodies works to reveal our own parts in the technological mediation of time and space, the proliferating contingencies in photographic mapping and the artworks that have reframed them for affective ends – fleeting, mundane, poignant, politically charged, sinister, uncanny, or just glitchy – surely work to expose, as never before, the map's silencing of its temporal codes.

## Notes

1  See https://goo.gl/maps/HEfpY (accessed 30 November 2017).
2  See http://realtime.waag.org/ (accessed 30 November 2017).
3  A survey of early artistic employments of GPS can be found in Stephen Wilson's (2002) *Information Arts: Intersections of Art, Science, and Technology*. Missing from that survey is Laura Kurgan's (1994) installation *You Are Here: Information Drift*, mentioned later in this chapter.
4  The criticality of 'locative media' (referring here to a genre of media art practice involving location-aware technologies and digital mapping) has been much debated (Albert, 2004; Fusco, 2004; Holmes, 2004; Tuters and Varnelis, 2006; Tuters, 2012).
5  Derrida's ideas about 'originary technicity' are returned to and developed throughout his oeuvre. Arthur Bradley (2011) has produced a useful, chapter-length review of them in his book on the concept.
6  Fujihata's GPS projects are documented at www.field-works.net/ (accessed 30 November 2017).
7  I am strongly influenced here by Bruno Latour's work on the 'metrological chains' through which spacing and timing regimes are produced (Latour, 1997).
8  Denis Cosgrove has identified the 'whole earth' as one of two rival, if related, discourses of global unity that drew upon the same images, distinguishing it from an imperialist 'one world' discourse (Cosgrove, 2003).
9  *You Are Here: Installation Drift* was exhibited at the StoreFront for Art and Architecture, New York, in January, 1994. It dealt with the workings and military provenance of GPS, and visualised the 'drift' of Selective Availability (Kurgan, 1994). Kurgan has recently published a collection of essays about her projects (Kurgan, 2013).

10  See www.l00k.org/monochromelandscapes/monochrome-landscapes (accessed 10 May 2016).
11  The other colours were represented by the Arctic National Wildlife Reserve, the intersection of the Greenwich meridian with the equator in the Atlantic Ocean, and an old-growth tropical rain forest in Cameroon.
12  See www.mishkahenner.com/Feedlots (accessed 30 November 2017).
13  See http://9-eyes.com/ (accessed 30 November 2017).
14  I am not going to dwell on the distinction between augmented reality and geomedia – in terms of the examples discussed by Graham, Zook and Boulton, the terms can be treated as interchangeable.
15  See http://projects.justinblinder.com/Vacated (accessed 30 November 2017).

# References

Albert, S. (2004) Locative literacy. *Mute*, 1(28). [Online] Available at: www.metamute.org/editorial/articles/locative-literacy (accessed 3 August 2016).

Bradley, A. (2011) *Originary Technicity: The Theory of Technology from Marx to Derrida*. London: Palgrave MacMillan.

Castells, M. (1996) *The Rise of the Network Society*. Malden, Massachusetts: Blackwell.

Cosgrove, D. (2003) *Apollo's Eye: A Cartographic Genealogy of the Earth in the Western Imagination*. Baltimore, Maryland: Johns Hopkins University Press.

Crampton, J. (2009) Cartography: Maps 2.0. *Progress in Human Geography*, 33(1): pp. 91–100.

De Certeau, M. (1984) *The Practice of Everyday Life*. Berkeley, California: University of California Press.

Dodge, M. and Perkins, C. (2009) The 'view from nowhere'? Spatial politics and cultural significance of high-resolution satellite imagery. *Geoforum*, 40(4): pp. 497–501.

Dorrian, M. (2013) 'On Google Earth'. In: Dorrian, M. and Frédéric P. (eds) *Seeing From Above: The Aerial View in Visual Culture*. London: I. B. Tauris, pp. 290–307.

Farman, J. (2012) *Mobile Interface Theory: Embodied Space and Locative Media*. New York: Routledge.

Fusco, C. (2004) Questioning the frame: Thoughts about maps and spatial logic in the global present. *In These Times*. [Online] Available at: www.inthesetimes.com/article/1750/ (accessed 8 August 2016).

Graham, M., Zook, M. and Boulton, A. (2013) Augmented reality in urban places: Contested content and the duplicity of code. *Transactions of the Institute of British Geographers*, 38(3): pp. 464–479.

Hansen, M. B. N. (2005) Movement and memory: Intuition as virtualization in GPS art. *Modern Language Notes*, 120(5): pp. 1206–1225.

Harvey, D. (1989) *The Condition of Postmodernity: An Enquiry into the Origins of Cultural Change*. Oxford: Blackwell.

Hemment, D. (2006) Locative arts. *Leonardo*, 39(4): pp. 348–355.

Henner, M. (2011) *No Man's Land*. Manchester: Self published book.
Henner, M. (2012) *No Man's Land Volume II*. Manchester: Self published book.
Holmes, B. (2004) Drifting through the grid: Psychogeography and imperial infrastructure. *Springerin*. [Online] Available at: www.springerin.at/dyn/heft_text.php?text id=1523&lang=en (accessed 8 August 2016).
Kurgan, L. (1994) You are here: Information drift. *Assemblage*, 25: pp. 14–43.
Kurgan, L. (2013) *Close Up at a Distance: Mapping, Technology and Politics*. New York: Zone Books.
Lapenta, F. (2011) Geomedia: On location-based media, the changing status of collective image production and the emergence of social navigation systems. *Visual Studies*, 26(1): pp. 14–24.
Latour, B. (1997) Trains of thought: Piaget, formalism, and the fifth dimension. *Common Knowledge*, 6(3): pp. 170–191.
Mackenzie, A. (2002) *Transductions: Bodies and Machines at Speed*. London and New York: Continuum.
Massey, D. (2005) *For Space*. London: Sage Publications.
Mitchell, W. J. T., and Hansen, M. B. N. (2010) 'Time and space.' In: Mitchell, W. J. T. and Hansen, M. B. N. (eds) *Critical Terms for Media Studies*. Chicago, Illinois: The University of Chicago Press, pp. 101–116.
Moakley, P. (2012) Street View and beyond: Google's influence on photography. *Time LightBox*, Weblog, 16 December. [Online] Available at: http://time.com/55683/street-view-and-beyond-googles-influence-on-photography/ (accessed 4 December 2017).
November, V., Camacho-Hübner, E. and Latour, B. (2010) Entering a risky territory: Space in the age of digital navigation. *Environment and Planning D: Society and Space*, 28(4): pp. 581–599.
O'Rourke, K. (2013) *Walking and Mapping: Artists as Cartographers*. Cambridge, Massachusetts: Massachusetts Institute of Technology Press.
Pinder, D. (2013) Dis-locative arts: Mobile media and the politics of global positioning. *Continuum*, 27(4): pp. 523–541.
Simondon, G. (1980) *On the Mode of Existence of Technical Objects*. Translated by N. Mellamphy (unpublished). London, Ontario: University of Western Ontario. [Online] Available at: http://dephasage.ocular-witness.com/pdf/SimondonGilbert.OnTheModeOfExistence.pdf (accessed 15 August 2016).
Speed, C. (2011) Kissing and making up: Time, space and locative media. *Digital Creativity*, 22(4): pp. 235–246.
Stiegler, B. (1998) *Technics and Time, 1: The Fault of Epimetheus*. Stanford, California: Stanford University Press.
Thielmann, T. (2010) Locative media and mediated localities. *Aether: The Journal of Media Geography*, 5(A): pp. 1–17.
Thrift, N. (2004) Remembering the technological unconscious by foregrounding knowleges of position. *Environment and Planning D: Society and Space*, 22(1): pp. 175–190.

Thrift, N. (2011) Lifeworld Inc. – and what to do about it. *Environment and Planning D: Society and Space*, 29(1): pp. 5–26.

Tuters, M. (2012) From mannerist situationism to situated media. *Convergence: The International Journal of Research into New Media Technologies*, 118(3): pp. 267–282.

Tuters, M, and Varnelis, K. (2006) Beyond locative media: Giving shape to the Internet of Things. *Leonardo*, 39(4): pp. 357–363.

Virilio, P. (1997) *Open Sky*. London: Verso.

Wilson, S. (2002) *Information Arts: Intersections of Art, Science, and Technology*. Cambridge, Massachusetts: Massachusetts Institute of Technology Press.

Wood, D. and Fels, J. (1992) *The Power of Maps*. New York: Guilford Press.

Zeffiro, A. (2012) A location of one's own: A genealogy of locative media. *Convergence: The International Journal of Research into New Media Technologies*, 18(3): pp. 249–266.

Zylinska, J. (2013) 'All the world's a camera: Notes on non-human photography'. In: Wombell, P. (ed.) *Drone: The Automated Image*. Bielefield: Kerber Verlag, pp. 162–174.

# 7
# Digital maps and anchored time: the case for practice theory

*Matthew Hanchard*

## Introduction

Digital maps are increasingly embedded within everyday practices, from choosing a holiday destination to gaining directions to a bar. As hypermediate and remediate forms (Bolter and Grusin, 2000), they are situated within a complex array of connected technologies: web mapping services output digital cartography via popular web map engines like Google and Bing Maps which, in turn, sit embedded on websites. Meanwhile, location-based services allow users to check in almost anywhere on the planet – volunteering their geolocation for public viewing on social media. Likewise, even seemingly unrelated practices like buying a house (landed capital investment) are now informed by digital maps. Property searches offer ready spatialisation of public datasets (school reports, crime statistics and boundary areas) set against the property type. Homebuyers now have the ability to narrow their shortlist criteria and create their own mapping prior to viewing, destabilising the sales practices of estate agents.

Alongside complex developments in the technological configurations of digital maps, and the entanglement with social practices, digital maps are increasingly ubiquitous through a complicated range of possible media. At times, this can negate meaningful analysis of digital map use through data alone: a digital map can be printed out and shoved in a back pocket, committed to memory, used as a back-up resource (just in case), or used in combination with a guide book or local knowledge. In turn, there is increasing complexity and challenge in grappling empirically with digital technology use beyond online-only web science.

This chapter argues that a practice theory approach, centring on how digital maps are used in everyday life, can contribute to the cartographic repertoire. Beginning with a sketch of cartographic theory from academic cartography to date, discussion places contemporary cartographic theory in context. This sets the scene in order to identify a historical limitation in cartographic theory that a practice theory of digital maps could address; namely, the wider anchoring of social practices. The following section provides an overview of practice theory for reference, outlining a simplified framework for analysis. The substantive section applies the framework, drawing on in-depth interview extracts. Three subtly altered socio-temporal practices are discussed through the lens of practice theory: maps as memory, organising time and altered routines. This serves to illustrate how a practice theory might be applied, in order to demonstrate the value it may add. In conclusion, I argue that a practice theory approach provides a useful means to address the relationship between digital map use and shifting temporalities in everyday life.

## The contours of cartographic theory sketched out

In this section, cartographic theory is laid out in brief to provide a context against which to situate the value of a practice theory of digital maps. Full histories of cartographic theory have been written elsewhere, with Perkins (2003), Edney (2005) and Crampton (2010) each providing excellent overviews. In summary, cartography has existed for millennia, and in various forms. However, cartographic theory emerged only immediately prior to the Second World War, largely through the work of Arthur Robinson (see, for example, Robinson, 1979) and surrounding researchers (Perkins, 2003; Edney, 2005). As the first sustained attempt to create a more accurate map, Robinson (1979) tried to incorporate a reflexive sensitivity towards end-users while working within boundaries of positivist normal science. Radical cartographers, map artists, map propaganda and psychogeography presented minor challenges (Crampton and Krygier, 2006), but cartographic practices remained relatively stable, even throughout the quantitative turn in geography (Robinson, 1979: 101). In the late 1970s and early 1980s, amidst a wider interpretive turn in social theory and emergence of critical, humanist and radical geographies (Dorling and Fairbairn, 1997: 142–145), the new field of critical cartography began to challenge the dominant perspective of academic cartography (Crampton and Krygier, 2006). This was achievable, first, through critical action-based research seeking to develop alternative representations (Bhagat and Mogul, 2008; Crampton, 2010) and psychogeographic mappings, drawing on a connection between the Avant-Garde and Situationist movements (Rasmussen, 2004); and second, through theoretical critique of the power relations between map content and spatial knowledge(s). On the latter, key moments include Harley and Woodward's *History of Cartography* (Andrews, 2001) – a massively ambitious (and on-going) project, intended to redress subaltern dynamics within map representation (Harley, 1987). In drawing on Harley's combination of post-structuralism, semiotics and social constructionism, the project sought to critique knowledge-politics in map representation (1988a; 1988b; 1989); and to challenge Wood's (1992; 2010) focus on power relations in mapping processes, and later, Black's (2002) focus on the embedded politics within map representation. Throughout the 1990s and 2000s, sophisticated Geographical Information Systems (GIS) emerged through technological innovation (especially in computing). Subsequent engagement between GIS practitioners (so-called 'GISers') and critical cartographers generated useful debates throughout the 'GIS Wars' (Pickles, 1995; Schuurman, 2000; Goodchild, 2006; O'Sullivan, 2006) – a term reflective of the wider 'Science Wars' (Flyvbjerg, 2001: 1). These culminated in several new fields informed by critical geography: critical GIS (Schuurman, 2008; Crampton, 2010); Public Participation GIS – often termed 'P/PGIS' (Craig, Harris and Weiner, 2002; Dunn, 2007); feminist GIS (Pavlovskaya and Martin, 2007; Elwood, 2008; Kwan, 2010) and postcolonial GIS (Crampton, 2010). Debate continues between GISci (GIS Science), directed towards specialist use of technical tools within a normal science position in line with academic cartography, and more humanist leaning neo-geographers (Turner, 2006) – the latter often opting to use web-based, publicly open proprietary web mapping applications, and informed by a constructivist stance. For example, using Google Maps and Bing Maps alongside 'grassroots' maps (Turner, 2006) to create inductively generated maps through end-user contributions, e.g. OpenStreetMap, inclusive of multiple perspectives and subjectivities.

Recent advances in processing and software technologies afford sophisticated cross-platform digital maps, ranging from popular web maps, e.g. Google Maps (including the realist Street View function) and associated API (Application Programming Interface), through to nation-specific proprietary remediation, e.g. Britain's Ordnance Survey (OS), provided via third-party vendors initially and then online. Alongside map-specific developments, innovations in related technologies continually afford increasing accessibility, ubiquity and mobility. The development of Ajax (Asynchronous Javascript and XML) alongside other Rich Internet Applications (RIAs) enables easy embedding of content within a webpage without the need to reload a page, or re-run search queries (Ying and Miller, 2013). In turn, this made on-screen digital map use far smoother for end-users, especially mobile users with low signal strength or bandwidth. Similarly, the standard shift from HTML 4 to HTML 5 in 2014 enabled more control and further ease at embedding video and audio content onto webpages, alongside better graphic vectors – facilitating faster and smoother scaling of digital maps (zooming in and zooming out) and richer content to be interconnected, e.g. hyperlinks to videos direct via the HTML5 video element. In turn, this enables semi-skilled web designers to integrate geolocation and geofencing capability onto cross-platform websites with ease (single webpage design for smartphone, tablet and laptop), including bespoke map layers for clients, and all achieved through a user-friendly digital map engine interface.

Despite the increasing entanglement and ubiquity of digital and mobile media in everyday life (Urry, 2008; 2010; Castells, 2009), there is a limited amount of focus on digital map use from a sociological perspective. Research on digital maps is often technocentric (and at times technologically determinist), centring on location-based service modelling, or spatial analyses. This is overt, in the form of usability or User Experience (UX) research (University College London, 2012) for example – a field sympathetic to GISci and academic cartography. At other times, this can be diffuse. For example, Sui and Goodchild (2011) focus on GIS in convergence with social media, remaining at a general level. They do not explore the more widely produced and consumed (prosumed) web-based digital maps. Instead, they opt for the now discontinued Google Latitude feature, and not Google Maps (or digital maps) holistically. As human geographers, their research is vital in understanding a specific configuration of technologies, but it leaves vital sociological questions unaddressed. In abstracting technology from social relations as the site of study, they shift focus away from the wider relationship between digital map use, and the mundane, quotidian, socio-temporal practices that make up everyday life.

A small subfield of dispersed cartographic theorists have begun to engage digital maps, with the most comprehensive landmark text arguably remaining *Rethinking Maps* (Dodge, Kitchin and Perkins, 2009a). This extends the initial impetus of critical cartography, to draw together a dispersed set of contemporary theoretical strands, summarised in 'a manifesto for map studies for the coming decade' (Dodge, Perkins, and Kitchin, 2009b: 220). The authors assert contemporary cartographic theory should attend to five key lines of inquiry – which crosscut and intersect. These five key lines of inquiry are as follows: first, the *interfaces* encountered, akin to screen-spaces. Second, a turn to *algorithms* opening up the black-box of map technologies to critique. Third, *cultures of map use*, drawing on visual and comparative media studies (including software/computer game studies) to engage with contextually localised uses. Fourth, *authorship* to explore altered power relations inherent within map production (including new prosumer affordances); and fifth, research on *infrastructure* focuses on the materialities of digital maps, both to 'consider the infrastructure that makes that make mapping possible' and to 'analyse the ways in which mapping modes contribute to infrastructure themselves' (Dodge, Perkins, and Kitchin, 2009b: 228).

While their five lines of inquiry cover the materiality of digital maps, circuits of production and power relations, they leave a need to expand on how digital map use is entangled, and how it anchors (and is anchored by) other mundane socio-temporal practices in everyday life. Recently, a few researchers have started to address this gap, primarily along the 'cultures of use' line of inquiry. For example, Brown and Laurier's (2005) ethnomethodological conversation analysis of map use in car journeys explores how map use is entangled with other driving related practices, but brackets out other social relations. Likewise, Perkins (2008) provides the most promising purchase through ethnographic research, drawing on actor-network theory (ANT) to explore mapping constructions and circuits of capital bounded within the localised contexts of specific case studies of specialised map use. Meanwhile, Hind and Gekker (2014) focus on moments of (social) interaction between user and technology interfaces (driver-car assemblages), through ludic interactivity, drawing out links to gamification, prosumption and networked individualism in car driving practices (social navigation).

To date, there are few sociological analyses that approach digital map use beyond specialised activity; that is, as just one mundane everyday practice entangled and embedded within many others. In part, this is a historical effect, where academic cartography focused on map design, early critical cartography focused in response on a rebuttal of positivism. It addressed the map as text and the embedded politics. With contemporary cartographic theory following on from critical cartography, focus lies largely on: positivist (and post-positivist) map making and design (UX/GISci); action-based research aiming to redress power imbalances (P/PGIS, neogeography); or theory that addresses cartography as a specific activity – the latter often bracketing out the wider social relations in which use is entangled, contra Perkins' (2008) position, on which this chapter builds. This is an important limitation for contemporary cartographic theory; current approaches are less amenable to enable understanding or theorising the ways in which digital maps can and do reshape our social world, without which the value of cartographic theory as a wider project cannot be asserted other than as a narrow, specialised area of interest.

In the interest of methodological pluralism, a practice theory of digital map use can contribute towards an understanding of the socio-temporalities involved with digital map use from this wider sociological perspective, employing a different lens. This wider perspective affords an understanding of how digital map use is connected to, and situated within, a range of complex digital-social arrangements.

## Practice theory: a simplified sketch

As a well-established field of study, practice theory can boast an extremely diverse set of influences – from the philosophies of Wittgenstein, Heidegger, Dreyfus and Taylor (Schatzki, 2001) to Merleau-Ponty (Couldry, 2004), and on through to the social (and cultural) theories of Bourdieu (1977; 1992), de Certeau (1988), and Giddens (1984). Ontologically, the position reconciles an older structure/agency debate, holding that 'the social is a field of embodied, materially interwoven practices centrally organised around shared practical understandings' (Schatzki, 2001: 3). This requires a separation between actions (individual performances) and practices. Here, 'practice' as a general term describes all human action (drawing on the German *Praxis*), and 'practices' describes complex series of embodied ways of doing and knowing, drawing on the German term *Praktiken* (Reckwitz, 2002). '[I]f practices are the site of the social – then routinized bodily performances are the site of the social and – so to speak – of social order' (Reckwitz, 2002: 251). This is not to say all practices are habitual routine; a distinction Henri Lefebvre made explicit in separating (while retaining dialectical relations between) cyclical and linear time (Lefebvre, 2004; Gardiner, 2012: 43).

Practice theory is not blind to post-humanist challenges either; see Knorr-Cetina (2001) on objectual practices, for example. Practices are materially mediated, and focus on foregrounded human practices (opposed to a rule of general symmetry) to elicit understandings of how objects are constructed, how objectually mediated performances are (re)enacted, and made meaningful, and on how 'bodies and "activities" are constituted within practices' (Schatzki, 2001: 2). Collectively, practice theory coalesces around agreement that shared knowledge is processual, tacit and embodied. This resonates with Geertz's 'thick description' of publicly observable ritual performances (cultural anthropology), where human actions speak to the wider cultural frameworks that simultaneously enable and constrain pace (Giddens, 1984).

From a sociological orientation, practice theory has two 'waves' (Schatzki, 2001; Couldry, 2004; Bräuchler and Postill, 2010). The first, exemplified by Bourdieu (1977; 1992), de Certeau (1988) and Giddens (1984), sought to theoretically resolve agency/structure dualisms through embodied practical action – practices. The second (contemporary practice theory) is not a unified 'school' of thought, but a 'body of highly diverse writings by thinkers who adopt a loosely defined "practice approach"' (Bräuchler and Postill, 2010: 7). For Shove, Pantzar and Watson (2012), practice theory can be summarised in a simple framework. Drawing on 'Innovation Studies' and 'Science and Technology Studies', they assert that technologies (such as digital maps) emerge as historically informed artefacts enacted within practices. They work through a combination of *materials*, *competences* and *meanings*. Where innovation occurs, new material forms are presented, e.g. digital maps present challenges to paper ones. Material forms require user competence for enactment (uptake of digital maps requires familiarity with a computer or mobile phone app, an understanding of paper maps or how maps operate, and access to relevant software). Similarly, competence and material form do not predicate change without a shift in meaning. Digital maps must be *understood* in some way as different from paper maps. This aligns with both Abrams (1980; 1982) and Wessels (2014) that social change is processual and historically informed.

For Shove, Pantzar and Watson (2012), focus lies on materials, competencies and meanings through two analytical categories – *elements* and *linkages*. 'Elements' are the pre-existing aspects that make up practices (the materials, competencies and meanings) considered as three simultaneously practically enacted parts. 'Linkages' are the connections that hold elements in place as they are practised. Any new practices may incorporate a change in elements, while shifts in linkages provide analytical purchase.

Where innovations (such as digital maps) are brought about, or new material forms are developed, new skills or competencies are required and/or gained, and new meanings emerge. These three 'elements' as Shove, Pantzar and Watson (2012) term them, are what constitute a practice (materials, competences and meanings). Any intervention in the form of a new technology, or challenge to existing practices, may destabilise relationships or 'linkages' in the terminology of Shove, Pantzar and Watson (2012). As a process, social change requires a continual destabilisation of practices.

Drawing on Bourdieu's (1977; 1992; 2010) concepts of cultural capital, economic capital, social capital and *habitus* (loosely – internalised dispositions) – a pre-existing social order can be seen within practices through linkages between elements. In practice theory, rather than mapping out the network of actants, and thus flattening social ontology, the focus sits closer to a 'Social Shaping of Technology' (SST) derived approach, centred on how embodied human activity (practice) produces and maintains social order through this framework of materials, competencies and meanings. It also asserts that the rules and resources available to an individual may limit or afford specific interpretations (meanings), competence or access to materials (Giddens, 1984). In Bourdieu's (1977) terms, economic capital, cultural capital and social capital are all at play, and infuse the set of predispositions an individual may hold, shaping their past and future practices.

Employing practice theory to address the cultures of digital map use diverges from some lines of inquiry set by Dodge, Kitchen and Perkins (2009a), connecting well with others. In focusing on meanings and competencies, it moves beyond material *interfaces* encountered as objects, or connected semiotics defined by the researchers' position. Similarly, in opening up the black-box of map technologies to critique, the focus shifts from a technological determinist stance on algorithms as *a priori* – either in the form of a diffuse background source code of power or as self-contained data products. Likewise, understandings of *infrastructure* shift from benign affordances, to humanly enacted constructions. Supporting materialities are not taken for granted either, but understood as contingent on competencies and meanings. Likewise, *authorship* no longer conceals a systems-based process – encoded from map information by the sender, and decoded or interpreted by an audience (with a minor note on new prosumer affordances as feedback loops). The approach is most closely aligned with the *cultures of map use* explored by Perkins (2008). Where Perkins (2008) draws on ethnographic methods informed by ANT, he provides important detail on contextually localised practices within specific case studies. In drawing on ANT, Perkins maps out the relationships (network) between various human and non-human actants; providing analysis on how local cultures are enacted and performed and how they are constructed or circulated through a praxeological lens (Reckwitz, 2002). A shift to practice theory moves beyond the moment of enactment, the actants involved, moments of translation or the assembled network. Instead, the focus lies on historically informed human action (practices not practice) – what people do – with the practices carried out as constitutive of social order (Giddens, 1984) – and not the individuals or maps directly.

Extending practice theory, Swidler's (2001) proposal for a focus on bodily-inscribed (embodied) action is a useful addition. She avoids idealism-materialism dualisms to focus on practices simultaneously in two directions. First, a move 'up' from internalised *ideas* (and away from Weberian 'world images') or *meanings* – in the narrow sense of conscious ways of knowing, not as a challenge to the use of the term by Shove, Pantzar and Watson (2012). Swidler's (2001) position is commensurable with Bourdieu's opposition to methodological individualism. In complement, the second move is 'down', away from impersonal *discourses* (Swidler, 2001: 74–75) in the sense of semiotic codes, structures and uniform understanding or interpretation (a position commensurable with Bourdieu's criticism of Levi-Strauss' structuralism). In taking up Bourdieu's (1992: 25–26) call to move beyond an either/or subjectivism-objectivism duality, and in simultaneously shifting 'up' away from methodological individualism, and 'down' from structuralism, Swidler (2001) turns to the more empirically researchable – that is the Geertzian inspired notion of *practices* as publicly observable processes (Couldry, 2004: 41). For Swidler, a turn to practices enables theory to attend to the relative importance of practices towards anchoring others, and situations of anchored practice; where some may be ritual reproductions, others 'public ritual practices ... able to create and anchor new constitutive roles' (Swidler, 2001: 90). Here, practices provide a locus to access social order as constituted through practices, alongside when and how practices anchor or organise entire practice bundles (Schatzki, 2002) – and social order.

Relating this approach to the cultures of map use in everyday life, the mundane, quotidian, taken-for-granted senses of time (temporality) may be assessed in four ways. First, through a sensitivity towards practices as historically informed and unfolding, contingent on technological innovations as material mediations that are socio-culturally practised by subjects. Second, through an understanding of digital maps as material artefacts that require some level of competence and meaningful sense of how and what to do with them (Shove, Pantzar and Watson, 2012) – as *cultural* artefacts. Here, practices act as the observable nexus of elements – material, competence and meanings that situate digital maps as innovations which may/may not (de)stabilise linkages. Third, through an understanding that diversity in practices (including negative cases or non-use) can provide insight onto the cultural, economic and social capital involved with digital map use. Fourth, through a focus on digital map practices that can be mobilised to explore how digital map practices can/do alter socio-temporal practices of everyday life, that is – how digital map practices anchor and are anchored by other social practices. The benefits of moving up to a more general level, and assessing digital map use as one practice among many others, does have a price; practice theory is limited in any epistemic guidance offered, and lacks any clear means for assessing how practitioners are recruited; a key strength of ANT.

## Towards a basic practice theory of digital maps

In the previous sections, both cartographic theory and practice theory are discussed in abstract terms: first, to identify and historically situate a specific gap that a practice theory of digital maps might address; and second, to set out a simplified framework for doing so. This section seeks to illustrate one way to operationalise the approach. There are several ways to employ practice theory, with several extant works.[1] In this section, excerpts from in-depth interview transcripts are analysed through the simplified practice theory set out above. Three socio-temporal issues are explored: maps as memory; organising time; and altered routines. Analytically, they do not depict a full analysis. At best, they are impressionistic and reflect the application of practice theory in sketch form only. These are not intended to depict a full analysis, but simply serve to highlight how practice theory can open up understandings of the ways in which digital map use anchors, and is anchored by, other socio-temporal practices. In short, this section illustrates that a practice theory of digital maps might be a useful means to understand how digital maps reshape everyday life.

## Maps as memory

Digital maps provide an affordance for new ways of remembering and reconnecting with personal biography. In this example, Sarah, a recently widowed retiree spent a large portion of her life in Africa (several countries), raising her children there while working with her husband for a non-governmental organisation (NGO). She notes on trying to use Google Maps and Google Street View via a laptop, as an aid to memory, that:

> I couldn't get the … level of detail that I wanted, to go and see where we lived, because I did want to go and see where we lived. I wanted to be able to zoom in and see the church where we got married, and I wanted to be able to zoom in on the houses we lived in each of those three places … you can't see the level of detail you can here … there are some significant buildings where you think, 'Oh, I should be able to find that' – you know …
>
> [W]e thought 'which old building is that?' and then we zoomed in, and had a guess as to which building it could be, having to think back, and bearing in mind I only lived there a couple of years. It was the town hall … it had just been a focus for town-build superiority, but years ago it had been a British colony, things like that. But the quality is poor … they still haven't got a photograph of that 'quality' …
>
> … It's what's important to them I guess – the users. I mean there aren't that many people in Mbale who are going to be zooming in to see where they live on a street map in the way we do here. (Interview conducted by author with 'Sarah' (pseudonym), 1 November, 2014)

For Sarah, digital maps are understood as anchoring of personal memory and history, alongside acting as a public historical record. The disconnect is user-generated; poor quality equates directly to the practices of others, where more users would drive a better quality map. Clear links to the politics of representation and the uneven distribution of image quality are of interest, set against collective forms of memory (or purposeful omission) of colonial histories – an ideal line of inquiry for a critical cartography project, and perhaps a P/PGIS or neogeographic project. Likewise, for practice theorists, an onus might lie on whether digital map prosumers lack the materials (allocative resources/ economic capital) to access digital maps, either through screen technologies or lack of underlying infrastructure, or whether competence in doing so (cultural capital) differs across national contexts. More importantly, practice theorists consider how such elements are linked and enacted. Here, the strength of a practice theory is in the simultaneous accounting of the cultural frameworks in which practices are carried out, e.g. limitations in national internet access infrastructure, while also accounting for the individual actions that constitute those frameworks through recursively stabilising linkages, e.g. using digital maps. However, remaining with the specific socio-temporal case at hand, the focus is on how digital maps are practised as mundane everyday technology in this context. As a socio-temporal practice, memory (collective and biographical) may be anchored by digital map use in different ways. Possessing both the required materials and competence to use a digital map, Sarah attempted to use Google Maps to virtually visit a past place, with digital maps meaningfully affording this practice elsewhere. She found disappointment where a colonial legacy continues to anchor the technical architecture and public data of the country (Kenya), and in turn, the quality of digital maps available as a memory resource. In this sense, digital maps can be seen as complex forms of memory.

## *Organising time*

Beyond the broad entanglement within affective senses of time or temporalities described above, digital maps are used in very instrumental ways. For Emma, digital maps help bridge alternative life rhythms. As a teacher, her working or 'available' hours differ from those of her partner, affording different routines:

> usually on the tablet while I am sat there, having a browse while something is on the telly. Um, but then there will be, you know, he might text me saying 'have you seen this house … have a look at this one …' and I'll look it up, and it will be some of the times either me at work being bored, or me at home having a look and going 'ooh, another one's come on the market, have a look at this one'. (Interview conducted by author with 'Emma' (pseudonym), 12 April, 2015)

In search for a house to buy, she is free to search from homes all day in the summer period in which schools are closed, while her partner is at work. At other times, she is at home but working. In order to manage the search criteria, several strategies are at play. Socio-temporal practices mediated by various material objects overlap pace (Couldry, 2004); Emma is actively watching television while also speaking on the phone, and navigating a property search. Meanwhile, her partner, stealing time from work, akin to de Certeau's (1988) *la perruque*, is remediating property search returns into an SMS text. As a shared practice (coordinated activity), digital maps can be seen to reconfigure temporality in two ways here. First, through a complex arrangement of materials and competencies used as authoritative resources, Emma and her partner are able to coordinate property search activities at more convenient timings. In turn, this affords retention of free time for her and her partner when physically co-present. Second, Emma notes an ability to use digital maps at moments of boredom, in which digital maps act as a means to make efficient use of non-productive time at work for private gain. In short, digital map practices allow the meaning of time to be altered subtly. In property searching when choosing home, activity can be coordinated and co-presence is no longer required. Likewise, previously unproductive time can be regained or repurposed. In everyday life, digital maps afford new ways of managing time.

## Anchored routines

In everyday life, socio-temporal practices are materially mediated in complex arrangements. They constitute a total field of practices (Schatzki, 2001; 2002), the sum of all practices anchored together. Following just one practice (holiday-making), two excerpts are drawn out from in-depth interviews, and discussed comparatively through the lens of practice theory to draw out the process of anchoring (Swidler, 2001).

The first, a long quote from an interview with two Bed and Breakfast (B&B) owners (Mary and Michael) outlines how digital map practices are entangled within sets of social relations, and anchor their daily routines:

> when we were younger, before all this lot came along, and we used to go away, sort of a few books, and the first thing we used to find out was the place. So, then you go off for the day and you know where you're coming back, whereas now, because of SatNav and Google Maps and everything else, we can have guests coming in at any time. I mean we say a three o'clock check-in, but they can be ten/eleven at night, because they've gone off.
> 
> ... people don't, they check-in when it gets to, I mean obviously it's light tonight, so they just check-in when they feel like ... we used to do it, especially if you would find somewhere that was in the middle of nowhere – down a track in Cornwall or whatever, you would go off and find it first and go 'right, now remember that road'.
> 
> ... these days people don't need to, because they know it's easy to find ...

> ... yeah, 'booking.com' all of them, they can write a review on – whether it is through the tourist board, or through late rooms, or booking.com, however you book you can put a review on afterwards, but most guests that I talk to, mostly use TripAdvisor, just to double-check that the place is going to be okay when they get here. (Interview conducted by author with 'Mary' and 'Michael' (pseudonyms), 14 June, 2014)

In these fragments, Mary and Michael discuss digital technologies broadly, identifying linkages between various elements that have been destabilised and restabilised (Shove, Pantzar and Watson, 2012). Where new material elements, e.g. SatNav, websites and web-based digital maps, have become a mundane part of their customers' holiday-making practices, they have subsequently destabilised linkages in holiday-makers' previously set 'checking-in' times. In their account, geospatial media and digital technologies are recent innovations, increasingly adopted in everyday life (requiring user competence). The account assumes their visitors perceive these technologies as meaningfully understood as more efficient and accurate than previous tools. In combination, they interpret their customers' use of maps as attributable towards an increased sense of ontological security (not getting lost), and an increased set of affordances.

However, this is not uniform – Mary and Michael use the operative '... *can have guests coming in at any time ...*' to suggest uncertainty (instability in practices). This leans towards a suggestion that cultural and economic capital may differ between visitors. Not all have access to the relevant materials (the technological artefacts) or possess competence in using them. Others may have both but remain predisposed to adhere to the suggested three o'clock check-in deadline through previous education as a form of social capital (Bourdieu, 1977). Alternatively, at the interplay of authoritative resources and rules (Giddens, 1984), their visitors may understand the technology an meaningful for increasing a sense of ontological security. At a descriptive and individual level, Mary and Michael later discuss the disruption to their entangled habitual routines and everyday life family rhythms. They had previously enjoyed an evening glass of wine watching television together on most nights, part of their dream for retiring and taking over the B&B. Instead, an increase in late check-ins brought about a perceived need for one of them not to drink. As a crude and singular example, their account describes the customer practices of checking-in late as anchoring of their family life routines. What this example depicts is that a practice theory affords meaningful analysis of digital maps within a wider set of social relations. One limitation of practice theory (a key strength of actor-network theory) is the analysis of recruitment and translation – the processes by which using the new technologies and sharing a sensibility towards acceptability of checking-in late are carried out.

In comparison, David notes that small shift in the practice of locating a hotel through digital maps and associated (entangled) technologies, affords an increased sense of ontological security:

> it would be really weird to go somewhere, like I say, if I know that I'm going somewhere, I'm kind of like already on TripAdvisor for example, I figure out if I could be staying there or, where are we going to stay, then decisions are made based on, yeah, absolutely. I mean, I needed to find a hotel at the weekend, again, somewhere we hadn't been before and I just went on Kayak and said 'I need hotels near this place'. I didn't then say map them for me. Because I knew where I was going to be and I could see them in a proximity to me. (Interview conducted by author with 'David' (pseudonym), 6 October, 2013)

In his account, digital maps are not directly recognised as tools he uses. Instead, he uses web applications (apps) previously downloaded onto his smartphone, for which he has competence in using (Kayak and TripAdvisor), and understands as meaningfully accurate (real) through experience. He views the apps as separate entities, despite both returning results from a cross-platform, bespoke Google Maps Engine derived layer delivered using Ajax technology and HTML 5. Both return search results weighted by user-generated reviews (prosumption) volunteered by others. As a hypermediate, at-hand, allocative resource enacted as authoritative resource (Giddens, 1984), a digital map is mundanely immediate, yet assists in hotel choice. The timing of doing so is not necessarily pre-planned. The map spatialises hotels in proximity to a set location, requiring far less time to plan, and thus his planning practices are afforded far more mobility.

In the two examples above, digital map practices anchor socio-temporal practices (check-in time, hotel choice and planning routines). It may be tempting to follow-up on the network, to assess how (and if) the local taxi firms have adjusted their operating times and fares to adjust to later check-in times, or how these changes have translated to the local pub meal serving times. Likewise, it may be interesting to focus on web development choices behind specific apps (opening the black-box). Instead, practice theory operates at a more general register; looking to the embodied practices (the act of checking-in late) and on the underlying tacit knowledge of human actors that carry them out (using the SatNav, using TripAdvisor, using Google Maps – all as resources). This includes affective knowledge (internalised dispositions) akin to Bourdieu's *habitus* (Bourdieu, 2010). In these fragments, the B&B owners' shift in daily practices is of central interest, noting that a minor change in the complex arrangement of technological and social practices of their visitors can have profound effects on their own domestic routines. The shift in planning and choosing a hotel is of interest too, but the analysis does not map out connections between the two through technological actants, or trace connections. Instead, the focus sits on how socio-temporal routines are de/stabilised; that is, how digital map use informs other socio-temporal practices (tracing connections between the practices, not the actants), and on how practices are enacted to construct the cultural frameworks that re-inform (enable/constrain) future practices.

## Conclusion

This chapter began with a sketch of cartographic theory, starting with the emergence of academic cartography, working through to contemporary approaches. Key leanings were introduced. At present, the central statement in cartographic theory remains Dodge, Kitchin and Perkins (2009a) manifesto that cartographic theory continues to coalesce around. This provided opportunity to turn cartography outward, and to expand the field towards a focus on maps broadly. Some theorists have done so, centring on how digital maps relate to infrastructure, exploring how software code flows from, into and through maps to alter the flows of people and things in urban space. Others focus on issues of authorship in an age of prosumer affordance. An important line of inquiry set out by Dodge, Kitchin and Perkins (2009a) was cultures of use. To date, this avenue has been explored by only a handful of researchers, ranging from: ethnographic case studies of specialised map use in localised contexts, drawing on actor-network theory (Perkins, 2008); through to actor-network theories of locative media in driving practices (Hind and Gekker, 2014). This chapter stresses the need for a theory of everyday digital map practices that centres directly on digital map use as a mundane activity situated within a complex arrangement of other socio-practices. In doing so, a practice theory framework is put forward in simplified form, and then operationalised in three short examples. This chapter serves to highlight a limitation of cartographic theory in gaining purchase on how digital maps anchor everyday socio-temporal practices, or on how various social practices in turn can anchor understandings and uses of digital maps. That is, on how digital map use shapes everyday life, and how changes in everyday life routines shape digital map use. Both are vital avenues, theorisation of which can only serve to strengthen the argument for the importance of cartography theory, given the complex arrangement and entanglement of digital maps and associated technologies.

It is worth noting that in this chapter practice theory has been isolated as a unitary approach in order to illustrate the strength it offers. However, practice theory can easily be combined with other approaches.[2] What a practice theory can offer is the ability to connect individual action with larger cultural frameworks; for digital cartographic theory this provides an ability to understand how digital map use fits within everyday life. In late modernity, digital technologies sit as ubiquitous background media, ready at-hand, and seamlessly integrated into an increasingly mobile (Urry, 2010) and technologically mediated world (Castells, 2009). Digital map integration is such that use may be unconscious, amenable only as tacit knowledge. It is only through an exploration of digital map use situated within everyday practices that entangled anchoring of practices can be accessed, with minute, mundane activities used to explore both the enacted cultural frameworks that inform digital map use and the ways in which digital maps contribute towards social order.

## Notes

1. For a diverse range, see Pantzar and Shove's (2010) practice theory of Nordic walking, Shove *et al.*'s (2008) extended book on a practice theory of DIY and digital photography, Nicolini's (2011) discussion of telemedine, or Murphy and Patterson's (2011) study of motorcycling edgework.
2. See González (2013) for an example of a practice theory based approach (PBA) and actor-network theory (ANT) in combination.

## References

Abrams, P. (1980) History, sociology, historical sociology. *Past & Present*, 87(1): pp. 3–16.
Abrams, P. (1982) *Historical Sociology*. New York: Cornell University Press.
Andrews, J. (2001) 'Meaning knowledge and power in the map philosphy of J. B. Harley'. In: Laxton, P. (ed.) *The New Nature of Maps: Essays in the History of Cartography*. London: John Hopkins University press, pp. 1–32.
Bhagat, A. and Mogul, L. (2008) *An Atlas of Radical Cartography*. Los Angeles: Journal of Aesthetic and Protests Press.
Black, J. (2002) *Maps and Politics*. London: Reaktion Books Ltd.
Bolter, J. D. and Grusin, R. (2000) *Remediation: Understanding New Media*. Cambridge, Massachusetts: Massachusetts Institute of Technology Press.
Bourdieu, P. (1977) *Outline of a Theory of Practice*. Cambridge: Cambridge University Press.
Bourdieu, P. (1992) *The Logic of Practice*. Cambridge: Polity Press.
Bourdieu, P. (2010) *Distinction: A Social Critique of the Judgement of Taste*. Abingdon: Routledge.
Bräuchler, B. and Postill, J. (2010) *Theorising Media and Practice*. Oxford: Berghahn Books.
Brown, B. and Laurier, E. (2005) Maps and car journeys: An ethno-methodological approach. *Cartographica*, 40(3): pp. 17–33.
Castells, M. (2009) *Communication Power*. Oxford: Oxford University Press.

Couldry, N. (2004) Theorising media as practice. *Social Semiotics*, 14(2): pp. 115–132.
Craig, W., Harris, T. and Weiner, D. (2002) *Community Participation and Geographic Information Systems*. London: Taylor and Francis.
Crampton, J. (2010) *Mapping: A Critical Introduction to Cartography and GIS*. Oxford: Wiley-Blackwell.
Crampton, J. and Krygier, J. (2006) An introduction to critical cartography. *Cartography*, 4(1): pp. 11–33.
De Certeau, M. (1988) *The Practice of Everyday Life*. London: University of California Press.
Dodge, M., Kitchin, R. and Perkins, C. (eds) (2009a) *Rethinking Maps: New Frontiers in Cartographic Theory*. London: Routledge.
Dodge, M., Perkins, C. and Kitchin, R. (2009b) 'Mapping modes, methods and moments: A manifesto for map studies'. In: Dodge, M., Kitchin, R. and Perkins, C. (eds) *Rethinking Maps: New Frontiers in Cartographic Theory*. London: Routledge, pp. 220–243.
Dorling, D. and Fairbairn, D. (1997) *Mapping: Ways of Representing the World*. Harlow: Pearson Education.
Dunn, C. (2007) Participatory GIS a people's GIS? *Progress in Human Geography*, 31(5): pp. 616–637.
Edney, M. (2005) Putting 'cartography' into the history of cartography: Arthur H. Robinson, David Woodward, and the creation of a discipline. *Cartographic Perspectives*, 51(36): pp. 711–728.
Elwood, S. (2008) Volunteered geographic information: Future research directions motivated by critical, participatory, and feminist GIS. *GeoJournal*, 72(3–4): pp. 173–183.
Flyvbjerg, B. (2001) *Making Social Science Matter: Why Social Inquiry Fails and How It Can Succeed Again*. Cambridge: Cambridge University Press.
Gardiner, M. E. (2012) Henri Lefebvre and the 'sociology of boredom'. *Theory, Culture & Society*, 29(2): pp. 37–62.
Giddens, A. (1984) *The Constitution of Society*. Cambridge: Polity Press.
González, G. (2013) The use of actor-network theory and a practice-based approach to understand online community participation. Doctoral Dissertation, The University of Sheffield.
Goodchild, M. (2006) Commentary: GIScience ten years after Ground Truth. *Transactions in GIS*, 10(5): pp. 687–692.
Harley, B. J. (1987) 'The map and the development of the history of cartography'. In: Harley, B. and Woodward, D. (eds) *The History of Cartography: Volume 1: Cartography in Prehistoric, Ancient, and Medieval Europe and the Mediterranean*. London: University of Chicago Press, pp. 1–42.
Harley, B. J. (1988a) 'Maps, knowledge, power'. In: Laxton, P. (ed.) *The New Nature of Maps: Essays in the History of Cartography*. London: John Hopkins University Press, pp. 51–82.
Harley, B. J. (1988b) 'Silences and secrecy: The hidden agenda of cartography in early modern Europe'. In: Laxton, P. (ed.) *The New Nature of Maps: Essays in the History of Cartography*. London: John Hopkins University Press, pp. 83–108.

Harley, B. J. (1989) 'Deconstructing the map'. In: Laxton, P. (ed.) *The New Nature of Maps: Essays in the History of Cartography*. London: John Hopkins University Press, pp. 149–169.

Hind, S. and Gekker, A. (2014) Outsmarting traffic, together: Driving as social navigation. *Exchanges: The Warwick Research Journal*, 1(2). [Online] Available at: http://exchanges.warwick.ac.uk/exchanges/index.php/exchanges/article/view/29 (accessed 8 August 2016).

Knorr-Cetina, K. (2001) 'Objectual practice'. In: Schatzki, T. R., Knorr-Cetina, K. and Von Savigny, E. (eds) *The Practice Turn in Contemporary Theory*. London: Routledge, pp. 175–188.

Kwan, M. (2010) Feminist geography and GIS. *Gender, Place & Culture*, 9(3): pp. 261–262.

Lefebvre, H. (2004) *Rhythmanalysis: Space, Time and Everday Life*. Edited by Stuart Elden. London: Continuum.

Murphy, S. and Patterson, M. (2011) Motorcycling edgework: A practice theory perspective. *Journal of Marketing Management*, 27(13–14): pp. 1322–1340.

Nicolini, D. (2011) Practice as the site of knowing: Insights from the field of telemedicine. *Organization Science*, 22(3): pp. 602–620.

O'Sullivan, D. (2006) Geographical Information Science: Critical GIS. *Progress in Human Geography*, 30(6): pp. 783–791.

Pantzar, M. and Shove, E. (2010) Understanding innovation in practice: A discussion of the production and re-production of Nordic walking. *Technology Analysis & Strategic Management*, 22(4): pp. 447–461.

Pavlovskaya, M. and Martin, K. S. (2007) Feminism and Geographic Information Systems: From a missing object to a mapping subject. *Geography Compass*, 1(3): pp. 583–606.

Perkins, C. (2003) Cartography: Mapping theory. *Progress in Human Geography*, 27(3): pp. 341–351.

Perkins, C. (2008) Cultures of map use. *The Cartographic Journal*, 45(2): pp. 150–158.

Pickles, J. (1995) *Ground Truth: The Social Implications of Geographic Information Systems*. New York: The Guildford Press.

Rasmussen, M. (2004) Situationist international, surrealism, and the difficult fusion of art and politics. *Oxford Art Journal*, 27(3): pp. 365–387.

Reckwitz, A. (2002) Toward a theory of social practices: A development in culturalist theorizing. *European Journal of Social Theory*, 5(2): pp. 243–263.

Robinson, A. H. (1979) Geography and cartography: Then and now. *Annals of the Association of American Geographers*, 69(1): pp. 97–102.

Schatzki, T. (2001) 'Introduction: Practice theory'. In: Schatzki, T. R., Knorr-Cetina, K. and Von Savigny, E. (eds) *The Practice Turn in Contemporary Theory*. London: Routledge, pp. 1–14.

Schatzki, T. (2002) *The Site of the Social: A Philosophical Account of the Constitution of Social Life and Change*. University Park, Pensylvania: Pensylvania State University Press.

Schuurman, N. (2000) Trouble in the heartland: GIS and its critics in the 1990s. *Progress in Human Geography*, 24(4): pp. 569–590.

Schuurman, N. (2008) Annals of the Association of American Geographers. *Annals of the Association of American Geographers*, 96(4): pp. 726–739.

Shove, E., Pantzar, M. and Watson, M. (2012) *The Dynamics of Social Practice: Everday Life and How It Changes*. London: Sage Publications.

Shove, E., Watson, M., Hand, M. and Ingram, J. (2008) *The Design of Everyday Life*. Oxford: Berg.

Sui, D. and Goodchild, M. (2011) The convergence of GIS and social media: Challenges for GIScience. *International Journal of Geographical Information Science*, 25(11): pp. 1737–1748.

Swidler, A. (2001) 'What anchors cultural practices'. In: Schatzki, T. R., Knorr-Cetina, K. and Von Savigny, E. (eds) *The Practice Turn in Contemporary Theory*. London: Routledge, pp. 74–92.

Turner, A. (2006) *Introduction to Neogeography*. Cambridge, Massachusetts: O'Reilly Media.

University College London. (2012) Say 'Hello' to GEMMA. *UCL GEMMA Project Blog*. Weblog, 11 November. [Online] Available at: http://gemma.blogweb.casa.ucl.ac.uk/ (accessed 17 July 2015).

Urry, J. (2008) *Mobilities*. Cambridge: Polity Press.

Urry, J. (2010) Mobile sociology. *The British Journal of Sociology*, 61(Suppliment s1): pp. 347–366.

Wessels, B. (2014) *Exploring Social Change: Process and Context*. Basingstoke: Palgrave-Macmillan.

Wood, D. (1992) *The Power of Maps*. London: The Guildford Press.

Wood, D. (2010) *Rethinking the Power of Maps*. New York: Guildford Press.

Ying, M. and Miller, J. (2013) Refactoring Legacy AJAX applications to improve the efficiency of the data exchange component. *Journal of Systems and Software*, 86(1): pp. 72–88.

# Part III

(In)formalising

# 8

## Mapping the space of flows: considerations and consequences

*Thomas Sutherland*

For many years now, scholars in geography, as well as other areas of the social sciences, have mounted a sustained challenge to the traditional theorisations of mapping which, in the words of Nigel Thrift (1996: 7), 'claim to re-present some naturally present reality'. There is, as such, little need to further rehearse such debates.[1] At the same time though, what we do need is greater theoretical intervention into the practices of representation that are inherent within mapping, and the ideological precepts by which they are informed and conditioned. Especially in an age of geographic information systems – wherein lies an increasingly stark disparity between the visual appearance of the map itself on one hand, and the numerical data that it claims to represent on the other – the parameters within which such representations are *given*, and the socio-political consequences of such 'givenness' must be analysed with intense scrutiny. Digital mapping *gives* us a world through the binding of quantitative information to a set of representational categories: the question that needs to be asked more strenuously regards what other worlds might be possible.

In this chapter, I will be focusing upon one particular component of digital mapping: the notion of *flows*, and the way in which they provide a common, but not sufficiently scrutinised, representational category for digitised spatialisations of the (inherently temporal) movement of people, goods and data. In the past twenty or so years, to speak of *flow* in the same way that political economists of a prior generation might have spoken of *circulation* has become utterly commonplace – even banal – to the extent that to critique it might seem pedantic. But in fact what we face is a discourse, especially in relation to the processes of globalisation, that takes flow to be a natural and unproblematic way of describing the temporalities and mobilities of digital, networked capitalism. I wish to challenge this, demonstrating that flow is not simply a neutral category, but rather, is a historically contingent mode of representation and givenness. Luc Boltanski and Ève Chiapello (2007: 143) describe it as, 'an organicist conception of society as a living body irrigated by flows, whether material (communication routes or systems for distributing energy sources) or immaterial (financial flows, flows of information, or movements of symbolic diffusion)' – which although not without its uses, smooths over the breaks, disjoints and dissymmetries that mark the globalised economy, and risks naturalising and even ontologising the myth of capitalism as a process of endlessly fluid expansion. Although I have written previously on the topic of fluidity as a theoretical and philosophical concept (see Sutherland, 2013; 2014b), its role as a metaphor used to represent time (i.e. movement, change, becoming, etc.) in a spatialised form within geographic information systems must be examined further.

## Flow maps

The notion of the 'flow map' as a distinct form of thematic map or infographic is comparatively recent, and one that is in large part tied to the gradual computerisation (and hence digitisation) of practices that were once completed by hand. Its introduction marks a decisive reorientation in the goals and principles of mapping, and a turn away from the representation of space towards the *spatialisation of time*. As Paul Virilio (2006: 71) puts it, we are in the midst of a shift whereby 'knowing-power, or power-knowledge, is eliminated to the benefit of moving-power – in other words the study of tendencies, of flows'. From the earliest days of the practice, cartographers would devise maps that not only recorded locations, physical features and political boundaries, but also trade routes and road networks. Of course, in an age of ubiquitous mapping – facilitated in large part by freely available mapping services, such as Google Maps, accessible by smartphone – such a concept seems rather banal (road atlases, produced by motoring associations, being a mainstay in the second-half of the twentieth century long prior even to these aforementioned digital services). But without the tools or competency to accurately chart such spaces, these maps were far more scarce for most of human history.

The first clear example of such a route map is the *Tabula Peutingeriana*, the sixteenth-century copy of an original Roman map most likely dating to somewhere between the first and third centuries CE, which traces a vast road network across the known world, from Rome itself through large sections of Europe, Africa, the Middle East and Asia (see Figure 8.1). The extent to which this document

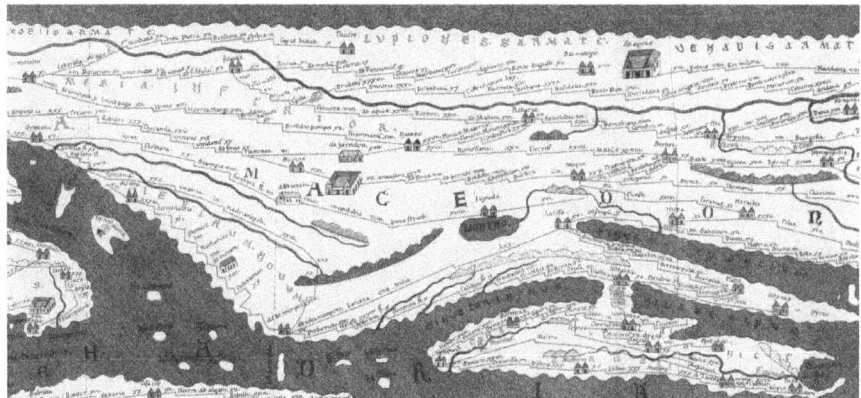

**Figure 8.1** Unknown author, *Tabula Peutingeriana*, c. fourth–fifth centuries. Conradi Millieri / Wikimedia / public domain.

might actually be considered a true map is contested, given the common suggestion that it is just a diagram of a route network visualising a pre-existing *itinerary* listing destinations along these roads.[2] But Richard Talbert (2010) argues that it does have a cartographic basis, even if highly abstract: while there are vast distortions necessitated in order to represent so much detail within such a narrow frame, this does not discount the amount of geographical information that is still represented. It is conceivable 'that no previous mapmaker had been so bold as to take a frame of such extreme dimensions and then to set the entire *orbis terrarum* within it, with the city of Rome as the center point – all of which required that the landscape be remolded on an epic scale' (Talbert, 2010: 162).

The design and accuracy of such route maps was greatly increased during the Age of Exploration (from the fifteenth until the eighteenth centuries), assisted by the various technical devices (e.g. compasses, telescopes, sextants, etc.) that allowed for more precise measurement of directions, angles, distances, and so on. These maps, which allowed the tracing of the vast courses travelled by traders, merchants and colonists across the globe, were particularly crucial for the expansion and management of the territory of the European imperial powers through to the twentieth century (see Figure 8.2). As we will see later in this chapter, such practices of mapping were also implicated in processes of epistemological rationalisation, essentially flattening out the spherical surface of the Earth so as to fix it within a predetermined (Cartesian) geometrical schema. What distinguishes this mapping of trade routes from the mapping of flows that we are discussing here is the former's fixity: such maps made no claims to recording the movements, directions and quantities of actual merchants or their goods; instead, they represent the paths that these objects *might* take. Most

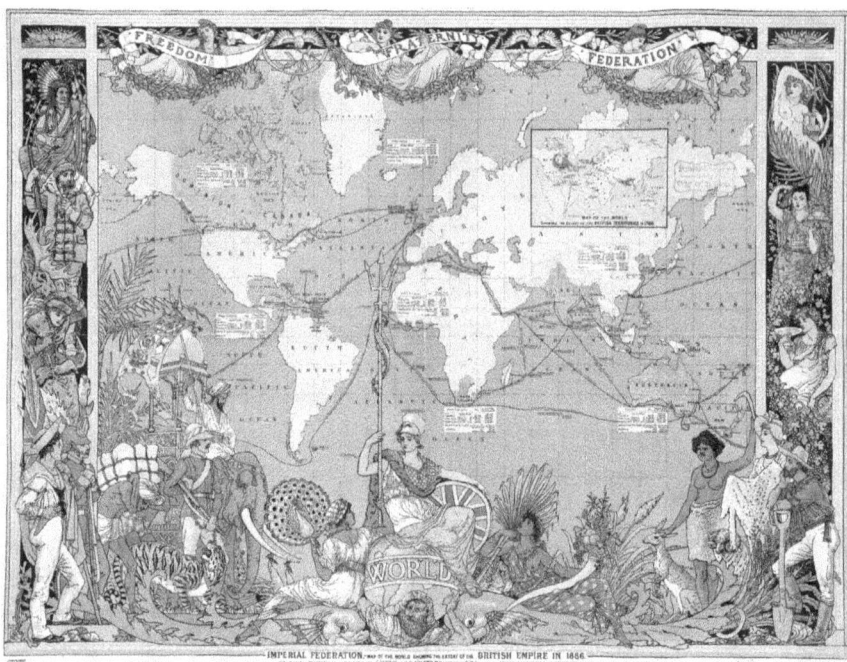

**Figure 8.2** J. C. R. Colomb, map of the British Empire from 1886. Norman B. Leventhal Map Center Collection, CC BY-NC-SA.

importantly, the temporal dimension is almost entirely absent, giving little or no sense of chronological change, and presuming the preservation of these routes over significant lengths of time.

Flow maps, by contrast are in essence an application of the principles of the flow chart to cartographic mapping (and more specifically to *thematic mapping*, wherein maps are designed to emphasise one particular subject area or concern). They are typically less interested in specific paths taken by actors, and more in what it is that is 'flowing', from where it originates and to where it is heading, the means by which it is moving, migrating or being transported, and the quantities, frequencies and velocities of this movement. They can be quite simple, indicating just an origin, a terminus and the direction of travel, or they can be very complex, illustrating multiple separate or divergent vectors, multiple directions, intermediaries through which flows pass, bifurcations and splits, transformations, varying quantities, speeds, capacities, and so on and so forth.

The first clearly recognisable examples of these flow maps all appear in the mid-nineteenth century, the key innovators in this respect being the British soldier and public servant Henry Dury Harness, British physician John Snow, and French civil engineer Charles Joseph Minard (see MacEachren, 1979). Harness, who worked for the newly founded Irish Railway Commission at the time, developed what is perhaps the first ever set of flow maps, representing both the movement of travellers and commodities, as well as the volume of such movements through the usage of lines of varying thickness. Snow, who had a distinguished career in medicine, attempted to dispute the hegemonic miasmatic theory of disease transmission by mapping the incidence of cholera across London, demonstrating that the infection was being spread via a water-pump, and was hence water-borne. Although this was not in itself a flow map as we would usually understand it, it nonetheless marks a significant milestone in both the rise of infographics as an apparently effective medium for visually representing information, and the increasingly ubiquitous use of statistics and other such instruments in order to survey and categorise a human population. Eugene Thacker (2004: 178) contends that 'bodies, though never apolitical, become politically materialised at the moment they are transmuted into policies, laws, governmental guidelines, funding sources, marketable and FDA-approved drugs, and medical-economic investments and insurances', and while it would not be fair to lump Snow's study in with such methods of control, it nonetheless signifies a crucial step towards such biopolitics.

Finally, Minard, who is renowned as a pioneer in the representation of numerical data through visualisation, and who is generally recognised as having developed the symbolisation of flows independently of Harness, produced a famed map charting Napoleon's disastrous march to Moscow in 1812 (see Figure 8.3). The failure of this endeavour marked the beginning of the decline of French hegemony in Europe. The map measured the diminution of the French army in terms of geographic location, as well as six other types of data, remarkably, in a single graph. It is probably this ability to represent so much information within a single image (often with decidedly social or political ends) that has led to the proliferation of the flow map in recent years. From Alexis Bhagat and Lize Mogel's *An Atlas of Radical Cartography* (2008), through the Bureau d'Etudes and their noted 'Governing by Networks', to the Counter-Cartographies Collective (3Cs), and the Spanish group Hackitectura, flow maps are of the moment. One project from Hackitectura epitomises this trend – seeking as it does to understand the border between Spain and Morocco as 'not an abstract geopolitical line but an increasingly complicated, contested space' by attempting to 'follow the flows that already traverse the border, such as migrants, internet data and cell phone calls, as well as capital and police', and the way in which these flows shape it into a border region (Dalton and Mason-Deese, 2012: 448).

As Robert L. Harris (2000: 157) writes, emphasising the diverse ranges of uses to which they might be put, '[f]low maps can be used to show movement

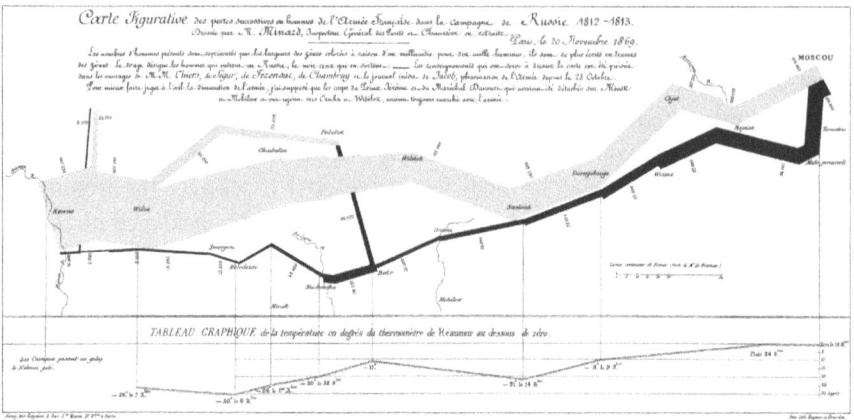

**Figure 8.3** Charles Minard, Carte Figurative, 1869. Mahahahaneapneap/Wikimedia Commons/public domain.

of almost anything, including tangible things such as people, products, produce, natural resources, weather, etc., as well as intangible things such as know-how, talent, credit, or goodwill'. Yet what is omitted in this description, and in most discussions related to this mode of mapping, is any justification for why such movement should be understood in terms of flow – a word which, from its Old English roots (*flōwan*) onward, is tied to the image of water in motion to a much greater extent even than the French *flux* (derived from the Latin *fluxus*). Such ambiguities date back at least as far as the Ephesian philosopher Heraclitus, whose metaphor of an endlessly flowing river is often (quite possibly erroneously) taken as an illustration of a universe in a state of constant 'becoming'. Does not the use of this metaphor project a certain set of *a priori* spatial qualities onto the temporal movements under question?

In response to this, one might point out that the term in this context is most likely derived from the commonly used concept of the flow chart, and its origins in the attempts to visually map out the problems of flow-shop scheduling, rather than any direct semantic linkage to the image of liquid flowing. While this is perhaps at least partly true, to make this claim would be to ignore the distance that its usage in mapping today has from these origins, *and* the way in which its usage has been shifted (and in some sense reciprocally determined) by its deployment across the social sciences, particularly but not exclusively in human geography. This notion of mapping flows, in other words, regardless of its semantic origins, fits quite neatly into a widely disseminated discourse that equates the metaphor of fluidity with, in the words of Zygmunt Bauman (2000: 5), 'the radical melting of the fetters and manacles rightly or wrongly suspected of limiting the individual freedom to choose and to act'. Flow is frequently associated with a series of global socio-economic transformations that have occurred in the second-half of the twentieth century. These include: trade liberalisation and the gradual elimination of tariffs, quotas and subsidies; deregulation of the financial and housing markets; the introduction of so-called 'flexible' labour practices, casualising the labour market and dramatically extending the precariousness of employment across varied industries; an increased corporeal mobility for both an elite managerial class, and a disenfranchised and displaced underclass; postmodernisation as the cultural logic of post-industrial capitalism, emphasising the mutability and non-essentiality of identity; and an economy that is ever more reliant upon the light-speed communication and transport of data through fibre-optic networks.

Of course, as Bernhard Siegert (2011: 14) notes, a media theory of mapping cannot understand the map as a mere representation in its own right; conversely, it should instead be 'concerned with the way changes in cartographic procedures give rise to various orders of representation', arguing that '[i]nstead of representing cultural predispositions', the map is 'their very basis of production'. To map trade routes in the fashion described above is not merely to re-present a pre-existing reality, formed in the shifts and manoeuvres of international commerce, for these maps themselves *give* a reality – one which provides possibilities, but also delimits them. Regardless though, the crucial thing to note is that route maps are indicative of a time when these possibilities were not expected to change with any great frequency or regularity. They can thus be understood as the products of an age when, in spite of the frenetic race to accumulation that marked imperialism and early industrial capitalism more broadly, time seemed to move more slowly and with greater predictability –this early period of globalisation stretched, rather than compressed, space and time, necessitating greatly deferred communication over vast distances.

It was only in the late twentieth century, contends Chris Speed (2011: 240), with the growing influence of human geography (itself in large part a reaction to the accelerating temporal milieu of neoliberal capitalism and increased demands for mobility engendered by an unprecedented push towards globalisation) that the mapping of space began to be really problematised by questions of time:

> [w]ith humans comes a model of 'time' that is more relative to 'real-time', and subsequently maps had to start speeding up. Formerly used to articulate the slow effects of ice-ages and other aspects of geo-morphology, maps now needed to show the speed of the human, and suddenly maps required time.

The complex relationship of this shift to processes of computerisation and the development of geographic information systems is particularly important here. It was through computer processing that complexity theory was able to move beyond the niches of mathematical modelling and become perhaps the dominant (i.e. hegemonic) paradigm for understanding change in the twenty-first century. As Manuel DeLanda (1991: 6) observes, computers have enabled the investigation of processes of self-organisation, whereby 'order emerges spontaneously out of chaos', the result being that 'natural phenomena once thought to lack any structure, like the turbulent flow of a fast-moving liquid, have now been found to possess an extremely intricate molecular organisation'.

In other words, even though for a long time flow maps were still predominately produced by hand, computerisation provides the means by which the mapping of self-organising systems over time might be understood in terms of 'flow'. In doing so, this also provides an imperative for the further mapping of such flows. I would propose that it is precisely this set of *technical conditions* that forms a necessary, albeit obviously in no way sufficient, cause for our (overdetermined) present day fascination with the representation of flows. Flow, anthropologist Stuart Alexander Rockefeller (2011: 557) argues, is 'one of the most important words constituting a new social scientific perspective on the relation of scale, agency, locality, and mobility on the global scene'. Yet, as he goes on to note, it is surprising how little it has been analysed given the import with which it is oft spoken, for '[t]he term has an aura and can appear to say a great deal, yet it can be employed in a nearly unaware fashion, as if its meaning were entirely uncomplicated and its use so innocuous as to call for no special mention', such that it might allude to quite radical implications while at the same time maintaining a certain etymological innocence (Rockefeller, 2011: 558). There is a seductive appeal – as well as a number of normative assumptions – contained in the words themselves and in the images that they evoke, and as such, there is no reason for us to treat them as neutral terms, nor ones whose meaning is unproblematically self-evident. In fact, it would seem to be their multivalent nature – the multiplicity of meanings for which they may be mobilised, often shifting between the parlance of metaphysics, natural science and everyday language without clear delineation – that has led to these terms' predominance.

The mapping of flows seems to take on a particular urgency in an age when the solidity and permanence of traditional socio-political structures upon which we have usually depended seems to be melting away in the furious creative destruction of neoliberal, globalised, digital capitalism. We might also connect it to changing patterns of representation engendered by new, ubiquitous forms of media – as Robert Hassan (2012: 179) writes:

> [t]he words we now interface with in social networking, in our news reading, in our working days and, above all, in our education are fluxual representations that

are mutable and flowing and no longer fixed in time and space as ink on paper is. Writing has become liquid, and digital representations of meaning have begun to pulse and flow at an ever-quickening pace that militates against the pause and traction, concentration and reflection that meaning construction and knowledge production demand, and that print culture could facilitate.

One is surely justified in wondering whether this emphasis upon the mapping of flows, rather than, or in addition to, routes and static locations, is indicative not only of shifts in socio-economic conditions, temporal environments and modes of data collection, but also the increasingly fluid means by which these changes are represented. These metaphors of fluidity tend to carry with them an implication of the affective or non-representational – in the words of Virilio (1994: 28), they mimic 'the gaze of the ancient mariner fleeing the non-refractive and non-directional surface of geometry for the open sea', seeking out 'environments of uneven transparency, sea and sky apparently without limits, the ideal of an essentially different, essentially singular world, as the initial foundation of the formation of meaning'. Yet are they not bound to those representations from which they hope to abscond through 'the paradoxical logic of the videoframe which privileges the accident, the surprise, over the durable substance of the message'? (Virilio, 1994: 65)

## Flow in the social sciences

Before we explore the specific consequences for digital mapping and geography, we should first explore further the use of the term flow within the social sciences, for it is here that the use of the category has largely become normalised. The term itself, being entirely mundane in origins, can be traced back almost to the beginnings of political economy as a field of study. Marx (1973: 211), for instance, speaks of 'the constant flow of the circulatory process' and political economists, in reference to mobility, use flow in terms of both goods and capital. Flow has also remained a standard term within the language of finance and the term first seems to have entered the social sciences as a distinct, albeit under-analysed concept in the work of Arjun Appadurai, an anthropologist whose work centres upon the interrelationships of globalisation and modernity in terms of a global cultural flow. For Appadurai (1996: 37):

> people, machinery, money, images, and ideas now follow increasingly nonisomorphic paths; of course, at all periods in human history, there have been some disjunctures in the flows of these things, but the sheer speed, scale, and volume of each of these flows is now so great that the disjunctures have become central to the politics of global culture.

While no adequate definition of the concept of flows is ever really provided at any point in his work, what would seem to be clear from this quote is that there is some sense of historicisation occurring. It is not so much the flows themselves (in the sense of freedom of mobility) that are new, but rather, the disjunction between the five dimensions (ethnoscapes, mediascapes, technoscapes, financescapes and ideoscapes) that comprise the social imaginary.

It would initially seem then that Appadurai's conception of fluidity is not singular or absolute, but instead defined by its turbulences and incommensurabilities. When he writes that the 'suffix *-scape*, allows us to point to the fluid, irregular shapes of these landscapes', it would appear that he is speaking solely in the language of contingency its turbulences and fluidity as a result of the speed and mobility that has been greatly facilitated by the processes of globalisation (Appadurai, 1996: 33). Simultaneously, however, Appadurai (1990: 301) relies quite heavily upon the distinctly metaphysical vocabulary of Gilles Deleuze and Félix Guattari, albeit with an anthropological tinge, when he argues, for instance, that '[d]eterritorialisation, in general, is one of the central forces of the modern world'. Deleuzian philosophy, it must be noted, has had a significant impact upon the ubiquity of this concept of flow within the social sciences and particularly human geography. In the words of Boltanski and Chiapello (2007: xxiv), what Deleuze and Guattari offer is 'an ontology containing only one tier or plane (the 'plane of immanence')', which 'knows only singularities or flows, the relationship between which assumes a reticular form and whose movements and relations are governed by a logic of forces'.

It is not particularly surprising that as a result, Appadurai (1996: 47) has a tendency to slip into an ontological register of writing: noting the importance of chaos theory, he speaks of a methodological approach premised upon 'a world of disjunctive global flows ... that relies on images of flow and uncertainty, hence *chaos*, rather than on older images of order, stability, and systematicness', and warns of naturalising 'the kind of illusion of order that we can no longer afford to impose on a world that is so transparently volatile'. The implication here seems fairly unambiguous: in an increasingly entropic social formation we can no longer justify preserving illusory concepts of order, and therefore, must embrace a methodological approach that embraces this chaos. For Appadurai, writes Rockefeller (2011: 561), 'flow is both the problem and the solution, the cause and the means of anthropological inquiry into globalisation, the reality that challenges our understanding and the tool to understand that reality'. The 'unyoking of imagination from place' – the deterritorialisation of imaginative power – he implies, should be embraced for its emancipatory potential (Appadurai, 1996: 58).

Manuel Castells' conception of flows, which he has developed from the late 1980s onward, is somewhat different from that of Appadurai, although they certainly share features. In *The Informational City* (1991: 169–170), he remarks that '[w]hile organisations are located in places, and their components are place-dependent, *the organisational logic is placeless*, being fundamentally dependent on the space of flows that characterizes information networks'. He expands upon this argument in *The Rise of the Network Society* (2010: 442), writing that:

> our society is constructed around flows: flows of capital, flows of information, flows of technology, flows of organisational interaction, flows of images, sounds, and symbols. Flows are not just one element of the social organization: they are the expression of processes *dominating* our economic, political, and symbolic life.

He defines these flows as 'purposeful, repetitive, programmable sequences of exchange and interaction between physically disjointed positions held by social actors in the economic, political, and symbolic structures of society', and posits them as constituting a *space of flows*, which he in turn defines as 'the material organisation of time-sharing social practices that work through flows' (Castells, 2010: 442).

What we see in Castells' work is an even more explicit historicisation of fluidity: he does not simply represent circulation and mobility as universal categories, but rather, reflects them in relation to a particular set of socio-technical and economic circumstances. The concept of the space of flows is also directly connected to the issue of temporality: it compresses time into a singular, homogeneous simultaneity – through increased speeds of computation, communication and data transmission, as well as the increased demands for multitasking and the dissolution of discrete social practices – producing an 'eternal ephemerality' in distinct contrast to the 'scattered, fragmented, and disconnected' temporality of the *space of places* (Castells, 2010: 497). What is profoundly valuable, though not unproblematic, about these accounts of flow is that they emphasise the enhanced role of the interconnected processes of circulation, distribution and transmission under digital capitalism (albeit in a specifically spatial form). Castells especially identifies with acuity the way in which demands for change and mobility are linked to the inhuman acceleration of the turnover time of capital.

Unlike Appadurai, Castells does not fall into the trap of celebrating the contingencies of flow, a rather common tendency which recalls, more than anything else, the mystical and irrationalist metaphysics of Henri Bergson (1911). Bergson (1911: 12, 46) counterposes an intuitional encounter with 'the flow of the real' against a rationalising, homogenising intellection which 'dislikes what is fluid, and solidifies everything it touches'. This equation of organisation and rationality, with oppression (and repression) exercising a significant, albeit often covert, influence over contemporary social theory, is achieved through his influence upon Deleuze. With this in mind, we may now look at the ways in which we might understand this notion of flows in relation to digital mapping and geographic information systems.

## The givenness of fluidity

In its simplest terms, digital mapping involves the digitisation of either pre-existing maps or the tracing and measuring of orthophotographic imagery. The latter refer to the use of geometrically corrected aerial photography as the basis for mapping, rather than using the traditional symbolic representations. Dissemination of imagery through services such as Google Maps has encouraged everyday utilisation of such data. In either case, the result is a set of spatio-temporally indexed digital data, which allows locations to be recorded in terms of both their physical placement (i.e. latitude, longitude and elevation, referenced through Cartesian coordinates) and, more crucially, their temporal occurrence. As a result, while traditional hardcopy maps or surveys will generally seek to represent spatial data at one particular point in time (or at most a few distinct periods, given that any more would likely make it uninterpretable), the digital map offers the possibility of representing and analysing changes over time with both minute detail and vast breadth. Hence, although the metaphor of flow as a figuration of capitalist circulation is not new, geographic information systems provide the capacity for a form of mapping *premised upon flows* – in other words, of tracing the specific movements of various diverse objects, patterns and events (e.g. not only people, animals, raw materials and commodities, but also, information, capital and affect, etc.) over a specified period of time, centralising the once marginal figure of time within these practices.

There is an inherent tension here, and one that is inevitable when discussing the temporal characteristics of mapping: to map is, in essence, to spatialise; to capture specific characteristics of the world within the fixed points of Cartesian space. This was of course the difficulty that Gerardus Mercator faced in 1569 when attempting to project a spherical globe onto a two-dimensional surface, achieving straight and perpendicular parallels and meridians, as well as a uniform linear scale, through the distortion of scale (see Figure 8.4). In effect, mapping necessitates distortion, a fact that becomes all the more obvious in the case of flow maps which are a *literal conversion of time into space*, inasmuch as they must somehow represent movement and chance according to the strictures of a spatial framework that makes no affordance to such temporalities. According

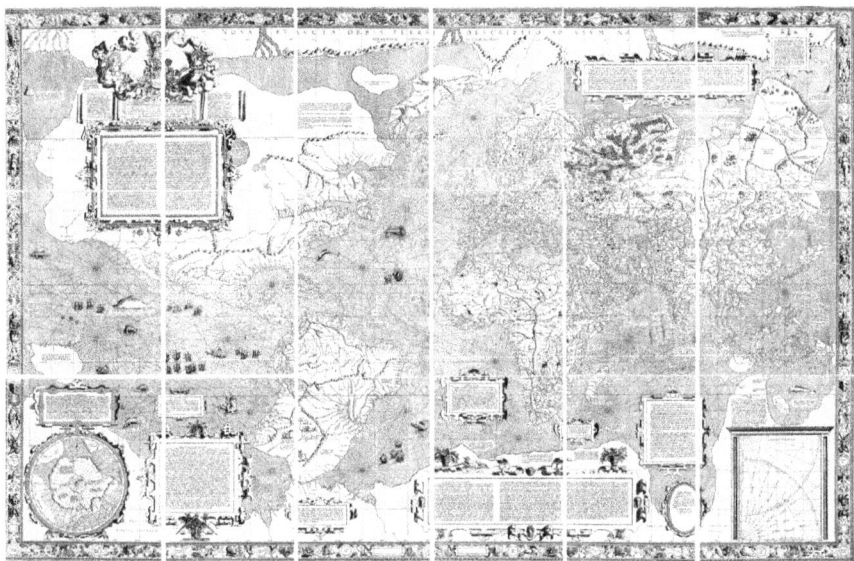

**Figure 8.4** Gerardus Mercator, world map, 1569. Alvesgaspar/Wikimedia Commons/ public domain.

to Cubitt (1998: 52), 'the conventions of traditional mapping embody ideological projects, portraying mastery over the environment, while the new GIS composite mapping techniques imply a similarly ideological domination over human geography'. Is there a specific ideology, we must wonder, attached to such processes of spatialisation (and the distortion of time therein)?

Bernard Stiegler (2011: 75) describes the various modalities of mediation, 'permitting symbolic fluxes and flows to be discretised and deposited, that is, permitting the spatialisation of their temporality', as processes of *grammatisation*. He argues that while such tendencies have always been a component of the technical mentality that defines human thought (insofar as it exteriorises itself), it is in the age of digital, networked media that grammatisation allows for the widespread exploitation of libidinal drives. Spatialisation, he suggests, is implicated in the rationalisation and subsequent homogenisation of human behaviour, giving it a quite specific ideological purpose. Of course, such anxieties regarding the transformation of time into space are not at all new: as far back as Aristotle's *Physics*, we see concerns from philosophers regarding the way in which time seems to be immobilised in its graphic representation. But what we face at present is a media environment in which our phenomenal experience of time is increasingly superseded, and indeed conditioned, by a growing set of digitally operated time-critical processes that are not only imperceptible (and perhaps even incomprehensible) to us, but which effectively *rationalise* the organisation and management of time according to principles that have little concern for human experience or wellbeing (see Sutherland, 2014a).

In this context, it is important to query whether the mapping of flows risks not so much representing the heterogeneity of temporal change as providing the means for its reduction to parameters advantageous to bureaucratic management and control. This is not to suggest that time is inherently opposed to spatial representation – an argument that would hew far too closely to the dualistic mysticism of Bergson or technophobic conservatism of Heidegger. Rather, the mapping of flows has become prevalent at a point in history when protocols of control, employed by both corporations and national governments, increasingly favour the management of time through its spatialisation and grammatisation. This does not mean that we should assume that there is a sinister character to flow maps, but it should make us aware of the potential that inheres within them for time (as it is represented) to be distorted along specifically ideological lines.

As noted earlier, one of the interesting elements of the flow map is its ability to graphically represent the movement or alteration of almost any object. One of the consequences of this broad applicability is that the practice of mapping flows has moved beyond the disciplinary boundaries of cartography and human geography to become a rather popular means of presenting information regarding movement and change throughout the social sciences. 'The flow map has an intriguing elegance', observes Mark Monmonier (1993: 190), such that 'the scholar with relevant data often cannot resist its ability to organise information and capture the reader's attention'. At a time when institutional, governmental and corporate pressures mean that academics within such fields find themselves needing to offer a veneer of objective scientificity to the research that they produce, even when this is grossly unsuited to their actual goals and methodologies, the mapping of flows provides a visually appealing and easily graspable way of fulfilling such expectations. It also risks *giving* a mode of representation (and in this sense then, giving a world in itself) that is structured in line with the exigencies of the rationalised, neoliberal academy. This is not to imply that other forms of mapping do not or would not comply with such exigencies in a similar manner, but merely to underscore the contingency of any particular instance of representation, embedded within the norms and conventions of its conditions of production.

When I speak of mapping as a mode of *givenness*, I mean specifically that a map is not simply a form of representation in its own right, but is a system through which a representation of the world is given. '[M]ap projections refer to their own systematicity', writes Cubitt (1998: 52), 'but also, as framing devices,

[they] point towards a world which they compose for the viewer but oppose to themselves'. When we speak in terms of the traditional map, it is not so much that an authorial subject (the cartographer) represents an objective eternal reality within which he or she resides, but that this precise division between subject and world is *given through* the process of mapping, establishing and refining the latter's contours and boundaries. Against a 'defiant insistence on a logic of representation, a common-sense belief in the evidence of an objective "reality" that is prior to all mental representations or written marks, a normative concept of rigour and scientism' (Siegert, 2011: 13), we can instead think mapping as a performative gesture productive of the phenomenality that we typically refer to as 'experience'. What makes digital mapping particularly worthy of commentary, however, is the way in which it quite visibly severs the presumed hierarchy between cartographer and world, situating instead in the latter's place the abstractions of digital data. Rather than making claim to the representation of an external, objective world, the procedural, generative and dynamic nature of digital mapping makes quite evident the way in which its production is reliant upon digitised, computerised data which *in no way resembles its supposed referent*. Luciana Parisi writes (2013: 18):

> generative algorithms are entering all logics of modeling – so much so that they now seem to be almost ubiquitous (from the modeling of urban infrastructures to the modeling of media networks, from the modeling of epidemics to the modeling of populations flows, work flows, and weather systems).

What digital mapping thus at least partly brings to light are the abstractions and ideals that lie at the heart of all cartographic practices, and which exist not as some necessary evil, corrupting and simplifying their referents in the name of utility, but as the very parameters through which this world is given. Whereas mapping in its traditional form was able to occlude this relationship because of the visual and spatial resemblance of its products to the empirical world that it claimed to represent, digital data and its generative algorithms are utterly incommensurable with analog visualisations that it produces. This quite starkly unveils the interfacial mediation of the digital. In the words of Alexander Galloway (2012: 82), data *'have no necessary visual form'*, and as such, require 'a contingent leap from the mode of the mathematical to the mode of the visual' in order to be represented. He goes on to remark that 'any visualisation of data must invent an artificial set of translation rules that convert abstract number to semiotic sign', pointing out that 'any data visualisation is first and foremost a visualisation *of the conversion rules themselves*, and only secondarily a visualisation of the raw data' (Galloway, 2012: 83). Once again, the process of representation that constitutes mapping is a process of givenness in its own right, rather than the mimesis of an already-given reality.

So how then does this relate to the notion of flows and fluidity? In the end, it is a question of critically interrogating the ways in which data are represented, given any visualisation in this respect is necessarily contingent (and thus arbitrary). We must not accept that any form of digital cartography is essential, natural or straightforwardly empirical (in the sense of a direct correspondence between representation and reality). What I wish to argue here is that the very concept of the mapping of flows, while potentially valuable in some instances, risks ontologising and absolutising the historical contingencies that it claims to represent, making them appear natural and unproblematic. In other words, the category of flows is a specific *form* of representation through which a distinctly, albeit not necessarily deliberately, ideological reality is given. In order to do this, I will return to the metaphor of flow as utilised in the social sciences, before reflecting upon its place within mapping more narrowly.

To say that '"everything flows", that matter is in flux, moving, becoming, is not to say that everything moves in the same way or at the same speed', contends Peter Merriman (2012: 5), for 'the world may be in constant movement, flux and becoming, but this does not mean that these movements are flat, linear and uniform'. In one sense, Merriman is correct – all things change, all things become, but they do so at their own rates, in relation to their own ontogenesis and patterns of individuation. At the same time though, is not the very notion of the 'flow' an exemplary case of the homogenising tendencies of such a discourse? Does not this metaphor carry along with it a set of affective, scientific and metaphysical presumptions that already shape and delimit these purportedly heterogeneous phenomena? It would be problematic to try to extricate the category of flow from the differential ontologies (those that identify being not with a stable, inert substance, but with the movement of a self-differentiating *becoming*) that have grown in influence over the course of the twentieth century. As Tim Cresswell (2006: 26) notes, whereas most traditional accounts of metaphysics (excluding perhaps the hydraulic model of Lucretian atomism) conceive of movement 'through the lens of place, rootedness, spatial order, and belonging', and thus view it 'as morally and ideologically suspect, a by-product of a world arranged through place and spatial order', this more recent form of ontology 'puts mobility first, has little time for notions of attachment to place, and revels in notions of flow, flux, and dynamism'. Perhaps such a category is needed, in order that these claims to heterogeneity might be given (and thus represented, within the strictures of theoretical discourse, if not the visual configurations of mapping), but this does not mean that flow is the *only* category that could be used, or that the connection between it and observations of movement or change is self-evident and unproblematic.

So why, we must ask, has this trope become so popular? Why is the term so frequently, uncritically and off-handedly deployed in the social sciences, and especially within the practices of mapping that have grown in dominance within these disciplines? The simple answer is probably to a large degree the correct one: the image of fluidity is an effective metaphor for the way in which network-driven distribution channels are able to transmit goods, information and even people at *rates and speeds* that make them effectively unthinkable by the human intellect alone, particularly when attempting to represent these movements in a visual manner. In one sense, the way in which digitisation has allowed such movements to instead be tracked through computerised algorithms – practical implementations of complexity theory – has allowed them to be characterised in this fashion, as a kind of simplistic formal cause: 'the way it tends to privilege a form (unbroken, agentless movement) over any content' (Rockefeller, 2011: 560). It does not matter what the contents of the flow are, as long as they flow, and as such, a contingency is raised to the status of a general category. Yet this is still an incredibly abstract sense of form, with no real recognition of the heterogeneity in speed, content and direction of these so-called flows. It is, in Alberto Toscano's (2008: 58) terms, a 'warm abstraction'. Far from the detached, static ideas that we usually associate with the abstract, he observes that 'recent conceptual production has sought to circumvent the customary reproaches against abstract thought by promoting concepts that are ever more vital, supple, pliant: flows, rhizomes, the virtual, scapes, the diagram, and so on' (Toscano, 2008: 58). In many cases, this term 'flow' appears to act as a floating signifier, used to describe some ineffable quality of the movements of a globalised world.

Marshall McLuhan (1964: 28) argues that 'the instant speed of electricity confers the mythic dimension on ordinary industrial and social action', and I wonder whether there is an element of this mythology in the conception of digital fluidity and its manifestation within the discourse of mapping: the seductiveness of the metaphor presents an effective way of mentally fathoming the overwhelming temporal complexity and acceleration of our world today. There is a worrying latent utopianism that seems to reside within this metaphor – writes Virilio (1994: 28):

> the power of the unexplored side of the failure of technical knowledge, a poetics of wandering, of the unexpected, the shipwreck which did not exist before the ship did; and beside this, very much alongside it, that stowaway, madness: the internal shipwreck of reason for which water, the fluid, remains a Utopian symbol throughout the centuries.

This is not to suggest that flow maps are necessarily embedded within anti-technological romanticism.³ Instead they can, regardless of the intentions of their creator, be read in such a fashion, and can simplistically depict a freedom of movement more in line with the mythology of digital capitalism than anything else.

None of this is to say that the mapping of flows is without utility, or that it is inherently politically reactionary. It is important that we elucidate the patterns of transnational commerce that define today's economy, especially given the way in which discourses of global development and growing labour markets belie the grossly unequal distribution of wealth and concentration of capital within a small set of post-industrial Western nations (even if this dynamic is gradually changing). Mapping flows also provides opportunities for understanding with greater clarity the ways in which specific types of commodities, particularly the products of informational and affective labour, are transmitted and distributed across borders and through diasporic communities, and can illustrate effectively the vast population movements and general processes of deterritorialisation that picked up speed during the twentieth century. There are a plethora of such opportunities available, and they should not be summarily dismissed. But at the same time, we must highlight that the metaphor of flows is not the only way to present such data, and is certainly in no sense a *natural* means of representation, having emerged from the very systems that we are attempting to critique.

There is a troubling level of obfuscation, as I hope to have shown, in this concept – as Rockefeller (2011: 564) notes quite accurately:

> certain usages of 'flow' carry some intellectual baggage that I doubt most people who use it would welcome – a radical time/space dualism and incompatibility with dialectical approaches. If we accept the terms of the dichotomy implicit in this genealogy of 'flow,' it becomes impossible to understand places or anything as the products of movement. Rather, things and movement remain in permanent opposition as appearance versus truth.

These categories of 'flow' and 'flows' are not neutral descriptors of the world, nor are they simply reductive and abstracted means of describing a reality that exceeds them. On the contrary, it is through these abstractions, at least in part, that such a distinction is given in the first place. Because data have no necessary visual form, we should not accede to any protestations regarding the inexorability of such representations. Flow, as a moulding of geolocational and temporal data into the form of a map, provides the representative frame through which the world is at least partly given, and is, in my contention, one that smooths over the disjunctures and dissymmetries that characterise the global economy today and problematise the mythology of a frictionless capitalism freed of any limitations or peripheries. The mapping of flows, in other words, even when it seeks to challenge the status quo, risks falling back into the same ideological practices of the state, military and financial institutions that, as Benjamin Noys (2010: 125) would have it, do not so much glide over an already-smooth space as 'constantly and actively *smooth* space' themselves.

Is it possible then to conceive of a politics of mapping, and a practice of representation that 'involves the preservation not merely of utopian moments or fantasies within the "smooth space" of capitalist ideology, but rather the memories and re-actualisations of forms and modes of struggle' (Noys, 2010: 169)? I would argue that it is, and that while we do not need to abandon the concept or application of the flow map (since this rejection would simply be an act of even greater obfuscation), we do need to augment or supplement it with a far greater level of attention to the breaks, disjunctures and striations that inhibit movements, as well as the institutional structures, both national and multinational, that coordinate them. Given how closely the very category of flow reinforces the mythologies of an entrenched and obdurate global capitalism, more effort should be made to indicate the inevitable disconnect between this and other potential modes of representation – that is, to show the contingency of this representation, and the gap between the givenness of the flow map and the data that is given (and hence shaped) through it.

Additionally though, what we might attempt to elucidate is the technical *mediation* that tends to be occluded in this form of presentation. Such mediation lies not only in terms of the various modes of computerised or otherwise mediated coordination that manage, direct and surveil these movements, but also resides in the technical conditions under which the processes of mapping occur, especially at a time when we are so heavily reliant upon geographic information systems and digitised processing of data. A denatured procedure of mapping flows would attempt to unveil such conditions of production, for as Wolfgang Ernst (2013: 52) writes, 'when a fiction is revealed, artificiality is also revealed, and the coming out of media is witnessed'. We should steer clear, however, of assuming that such a revelation is indicative of the fictional status of maps in relation to the objective reality that they supposedly represent; rather, by unveiling the mediation that lies at the heart of the very givenness of mapping as practice and decision, we can collapse this division. The point on which we should, and indeed must end then is that maps are always in some sense fictional, for they do not represent a reality as much as *give* the phenomena through which a reality is at least partly created. The question is, is the mapping of flows the most appropriate kind of fiction?

## Acknowledgements

Due to the limitations of the print format, some detail has been lost from the images contained in this chapter. For high-resolution colour versions, please see the Open Access edition at http://doi.org/10.9760/9781526122520.

## Notes

1 For a more detailed overview of the problems of representation in practices and discourses of mapping, see Del Casino and Hanna (2006).
2 A more contemporary equivalent of the *Tabula Peutingeriana* perhaps being the now-common public transport maps created in the wake of Henry C. Beck's 1933 circuit-like redesign of the London Underground's various lines: a remarkably clear but also highly abstract representation of a series of locations with little affordance made to either location or distance, as opposed to a map that is primarily grounded in a representation of physical space and the objects contained within it.
3 Plato is arguably the originator of this kind of thinking. In the *Protagoras* he equates the movement of water with the slipperiness of sophistry, counselling the titular character (one of the first sophists) to avoid sailing out into the open sea of false speeches and misleading rhetoric.

## References

Appadurai, A. (1990) Disjuncture and difference in the global cultural economy. *Theory, Culture and Society*, 7: pp. 295–310.
Appadurai, A. (1996) *Modernity at Large: Cultural Dimensions of Globalization*. Minneapolis, Minnesota: Minnesota University Press.
Bauman, Z. (2000) *Liquid Modernity*. Cambridge and Malden: Polity.
Bergson, H. (1911) *Creative Evolution*. Translated by A. Mitchell. Mineola: Dover.
Bhagat, A. and Mogel, L. (eds) (2008) *An Atlas of Radical Cartography*. Los Angeles: Journal of Aesthetics and Protest Press.
Boltanski, L. and Chiapello, E. (2007) *The New Spirit of Capitalism*. Translated by G. Elliott. London: Verso.
Castells, M. (1991) *The Informational City: Information Technology, Economic Restructuring, and the Urban-Regional Process*. Oxford: Blackwell.
Castells, M. (2010) *The Rise of the Network Society*. Chichester: Wiley-Blackwell.
Cresswell, T. (2006) *On the Move: Mobility in the Modern Western World*. London: Routledge.
Cubitt, S. (1998) *Digital Aesthetics*. London: Sage Publications.
Cubitt, S. (2013) 'Global media and archaeologies of network technologies'. In: Graves-

Brown, P., Harrison, R. and Piccini, A. (eds) *The Oxford Handbook of the Archaeology of the Contemporary World*. Oxford: Oxford University Press, pp. 135–148.

Dalton, C. and Mason-Deese, L. (2012) Counter (mapping) actions: Mapping as militant research. *ACME: An International E-Journal for Critical Geographies*, 11: pp. 439–466.

DeLanda, M. (1991) *War in the Age of Intelligent Machines*. New York: Zone Books.

Del Casino, V. J. and Hanna, S. P. (2006) 'Beyond the 'binaries': A methodological intervention for interrogating maps as representational practices. *ACME: An International E-Journal for Critical Geographies*, 4(1): pp. 34–56.

Deleuze, G., and Guattari, F. (1988) *A Thousand Plateaus: Capitalism and Schizophrenia*. London: Bloomsbury Publishing.

Ernst, W. (2013) *Digital Memory and the Archive*. Minneapolis, Minnesota: University of Minnesota Press.

Galloway, A. R. (2012) *The Interface Effect*. Cambridge: Polity.

Harris, R. L. (2000) *Information Graphics: A Comprehensive Illustrated Reference*. Oxford: Oxford University Press.

Hassan, R. (2012) The knowledge deficit: Liquid words as neo-liberal technologies. *International Journal of Media and Cultural Politics*, 8: pp. 175–191.

MacEachren, A. M. (1979) The evolution of thematic cartography: A research methodology and historical review. *The Canadian Cartographer*, 16: pp. 17–33.

Martin, E. (1996) The society of flows and the flows of culture: Reading Castells in the light of cultural accounts of the body, health and complex systems. *Critique of Anthropology*, 16: pp. 49–56.

Marx, K. (1973) *Grundrisse*. Translated by M. Nicolaus. London: Penguin.

McLuhan, M. (1964) *Understanding Media: The Extensions of Man*. New York McGraw Hill.

Merriman, P. (2012) *Mobility, Space, and Culture*. London: Routledge.

Monmonier, M. (1993) *Mapping It Out: Expository Cartography for the Humanities and Social Sciences*. Chicago, Illinois: The University of Chicago Press.

Noys, B. (2010) *The Persistence of the Negative: A Critique of Contemporary Continental Theory*. Edinburgh: Edinburgh University Press.

Parisi, L. (2013) *Contagious Architecture: Computation, Aesthetics, and Space*. Cambridge, Massachusetts: The Massachusetts Institute of Technology Press.

Rockefeller, S. A. (2011) Flow. *Current Anthropology*, 52: pp. 557–578.

Siegert, B. (2011) The map *is* the territory. *Radical Philosophy*, 169: pp. 13–16.

Speed, C. (2011) Kissing and making up: Time, space and locative media. *Digital Creativity*, 22: pp. 235–246.

Stiegler, B. (2011) *The Decadence of Industrial Democracies*. Translated by D. Ross and S. Arnold. Cambridge: Polity.

Sutherland, T. (2013) Liquid networks and the metaphysics of flux: Ontologies of flow in an age of speed and mobility. *Theory, Culture and Society*, 30: pp. 3–23.

Sutherland, T. (2014a) Getting nowhere fast: A teleological conception of socio-technical acceleration. *Time and Society*, 23: pp. 49–68.

Sutherland, T. (2014b) Intensive mobilities: figurations of the nomad in contemporary theory. *Environment and Planning D: Society and Space*, 32: pp. 935–950.

Talbert, R. J. A. (2010) *Rome's World: The Peutinger Map Reconsidered*. Cambridge: Cambridge University Press.

Thacker, E. (2004) *Biomedia*. Minneapolis, Minnesota: University of Minnesota Press.

Thrift, N. (1996) *Spatial Formations*. London: Sage Publications.

Toscano, A. (2008) The culture of abstraction. *Theory, Culture and Society*, 25: pp. 57–75.

Virilio, P. (1994) *The Vision Machine*. Translated by Julie Rose. Bloomington and Indianapolis: Indiana University Press.

Virilio, P. (2006) *Speed and Politics: An Essay on Dromology*. Translated by M. Polizzotti. Los Angeles: Semiotext(e).

# 9
# Maps as foams and the rheology of digital spatial media: a conceptual framework for considering mapping projects as they change over time

*Cate Turk*

## Introduction

> The world of mapping has rapidly moved from provisioning users with static two-dimensional hard copy displays to maps that are on-line, immediate and dynamic. (Cartwright, 2013: 56)

With a curious twist, we have come to think of a map like a 'folding' map that we carry around on our travels – a tactile three-dimensional thing with movement encapsulated in its title – as *static* as Abend also argues in this volume. This kind of idea contrasts with the flat-screen worlds of digital mapping at which we gaze (often while sitting relatively inert). William Cartwright (2013) refers to a transition in mapping that is happening in our time. Published paper maps that provide static depictions of places, frozen at the moment of compilation, are being replaced by digital mapping which enables dynamic, interactive visualisations where map readers can track changes or make changes over time. In this chapter I explore how this dynamism changes the way we think about and study mapping. Unravelling the curious twist, I consider how maps can be dynamic in a number of different senses.

I begin by examining two senses in which contemporary maps or 'new spatial media' (Elwood and Leszczynski, 2012: 544) are dynamic. In the first sense, 'dynamism' is due to the technological possibilities of these new media, such as 'slippy interactive mashup map objects'; and second a 'dynamism' is described by theoretical perspectives drawn from contemporary critical cartography which see maps as 'mutable' and 'manifold'. Turning to the questions this raises about how we might study maps as malleable changing objects, I suggest an analytical approach based on philosopher Peter Sloterdijk's (1998; 1999; 2004) concepts of *bubbles*, *spheres* and *foams*. These models, I argue, prove a useful way to conceptualise the fluid, contingent networks of relations that constitute dynamic mapping projects. Then, in order to illustrate how this conceptual model works, I examine a selection of crisis mapping practices focusing on the relations between maps and 'produsers'[1] (Bruns, 2006: 276); as well as the communities and interests that help determine the ultimate success and utility of crisis mapping efforts. Throughout, 'bubbles', 'foams' and the application of rheology (the physics of deformations and flows of matter) to mapping are up for discussion and critique.

## Dynamic maps and new spatial media

In the online environment, maps can be very obviously dynamic; with interactive visual interfaces, possibilities for inputs by multiple produsers, and continually accumulating datasets. Functions like 'slippy maps' (where the map moves within a screen revealing more and more territory as we scroll) enable a haptic engagement where maps shift beneath our fingertips. Coding protocols promote the ready mixing and mashing of data into map interfaces, and datasets may continually accumulate as new data is fed into a map over time (sometimes purportedly in 'real-time'). Moving to mobile devices, we see maps change through our movement, navigating with us.

There are at least four different ways in which these new 'digital spatial media interfaces' are dynamic. First, there are animated visualisations, where different map features move on a screen interface to provide more information. For example, the effective animation in Figure 9.1 shows winds as they move around the globe.

Second, we see dynamism as a result of user manipulation of the map interface, such as slippy maps or the ability to select a particular base map, perform a sorting of map layers and so forth (Figure 9.2).

With the use of maps on mobile devices, user movement is integrated into the dynamic map interface such that the map moves in ways according to user location or preferences.

Third, there are maps where datasets are continually or periodically updated, either through automated feeds, user-entered data or the addition of updated information by the map administrators. The combination of crowd-sourcing with mapping encourages members of the public to add their own data or change data. The updated content might be geographic features in a base map or the

Maps as foams 199

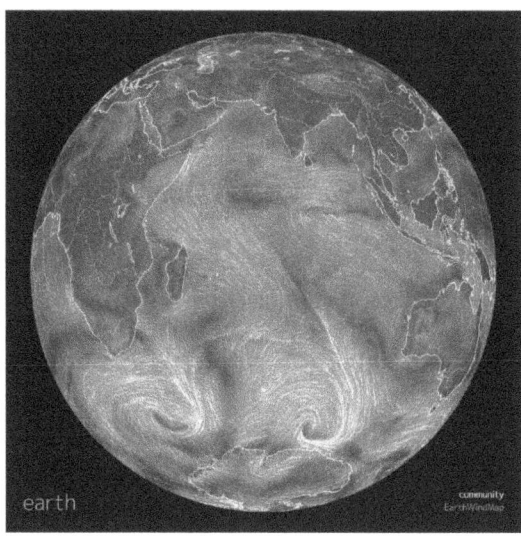

**Figure 9.1** Wind map, showing the way winds are flowing around the Earth (https://earth.nullschool.net).

**Figure 9.2** Haiyan/Yolanda Swipe Map, enabling comparison of before and after satellite imagery (www.esri.com/services/disaster-response/hurricanes/typhoon-haiyan-yolanda-swipe-map).

contribution of additional thematic data (see, for example ESRI, GIS Corps and StandbyTaskforce, 2013, where photos were added over time into a data layer).

The possibilities for recombination offered by digital presentation bring a fourth kind of dynamism. Maps are web elements which can be easily be cut, pasted, mashed-up, re-purposed and hyper-linked. Hence, the surrounding web context of the map changes. The website in which a map is embedded equates to what Woods and Fels (2008: 8–12) term the 'paramap', that is, the context surrounding a map that affects how it looks and is interpreted. The 'paramap' includes map elements such as title and legend (known as the 'perimap') but also extends to any documentation that explains what the map is showing (the 'epimap'). Maps thus change when a website design is updated or the map is re-presented in new sites, such as when a map is republished in a news report.

In each of these four ways, the content of the map and/or the interface is dynamic. Maps move, acquire data, have multiple authors and are adapted and re-purposed. The re-presentation of maps in a digital form also has implications for how we might 'date stamp' them or order them temporally. Websites might be periodically updated but there is no one moment of publication as there is with a paper map. Indeed, digital media in general is complex in terms of temporality. Following Tim Barker's writing about *Time and the Digital*, particularly his discussion of the philosophical work of Serres and Deleuze, it is probably useful to think of digital media as combining multiple temporalities (Barker, 2012). For example, when we access websites we consider them as 'present' or even as being in 'real-time', yet they build on a concatenation of data, conventions and actions from the past, and go on to have implications affecting users into the future. This complex temporality adds a further dimension to the dynamic map content and interactive interfaces described above. In the next section, I first discuss mutability of map objects and the implications of contingency for the study of maps or mapping projects, this leads on to a deeper consideration of temporality.

## Mutability and contingency

Various authors writing within sciences studies and critical cartography have explored the ways in which maps change and move. Bruno Latour (1990: 37) famously coined the term 'immutable mobile' to describe how knowledge (geographic and otherwise) moves through the world using transferable yet fixed ways of understanding or acting – see Latour (1990). Sybille Lammes (2011) draws on this when analysing the mutability of maps used in online gaming contexts. Lammes (2011) describes how due to the mobility of players and the mutability of the 'image of the map' (i.e. the graphic interface) the 'playing field' has become transformable, rather than static as with conventional board games. To Lammes, this makes the map almost a *'mutable* mobile' (Lammes, 2011: 3). But not quite, I presume, because there are still some enduring elements: conventions, code, rules and expectations that enable these maps to be recombined and reproduced. Nevertheless, Lammes (2011: 3) claims, 'the map itself has lost some of its immutability since the image of the map is constantly altered by the actions of the mobile player'. As such the visual experience of cartography within games, and much other digital spatial media, is dynamic.

Broadening our view to expand out from the central 'map image' to consider a mapping project or set of practices within which maps are embedded, we find another sense in which maps are dynamic, such that even good old paper maps, seemingly static on the page, are dynamic too. Del Casino and Hanna (2006: 36) call maps 'mobile subjects': 'infused with meaning through contested, complex, intertextual, and interrelated sets of socio-spatial practices'. They account for these practices in their study of the map use of tourism planners and tourists exploring place. Their work demonstrates that map making does not stop with the cartographer and continues through use and performative reproduction. The extension from 'map' to 'mapping project' signals the expansion of analysis away from the map object to a whole assemblage of actors – an ever-shifting constellation of the various cartographers, software, conventions, organisations and data sources (particularly previous maps), that, in combination, work to make, and continually re-make, a map.

Even when a map has been printed on paper fixed at a point in time, this network of actors and relations is fluid. The contingency of cartographic processes such as data collection, assigning of categories, and the different circumstances of map use (the same map being used in different ways in different circumstances) mean that maps as they are put together, reworked, folded or read are constantly in a state of becoming. Martin Dodge and Rob Kitchin (2007: 331) thus consider maps to be 'ontogenic'; their very being changing through use and context. There is a parallel here to Tim Barker's use of process philosophy in conceptualising digital media. He writes:

> the digital image, whether static or in motion, is the result of continuous and ongoing computations. It does not exist as a thing made but as a thing that is continually *in the making*. (Barker, 2012: 264)

Maps in whatever form are not static objects, but rather are dynamic, fluid and emerging.

The implications of such ontological instability for research about cartography are manifold. For critical cartographers, there is a practical conundrum of how to study and account for continually changing objects: if maps are so dynamic, continually changing, contingent upon context and use, how can we pin them down so as to discuss their content, intent and effects? Do we need to freeze these shifting multiple objects in order to analyse them? Through what means can we best understand these mobile subjects, mappings and mutabilities?

Several authors have approached this conundrum by taking what Kitchin, Gleeson and Dodge (2013: 480) describe as a 'processual approach'; analysing not only the visual content of maps, but the practices of production and consumption, performance and negotiation associated with maps or mapping projects. In practical terms, a variety of methods are used to examine mapping processes. Kitchin, Gleeson and Dodge (2013) employ an insider ethnography to relate the dynamic process of data collection and map making/use/re-use/re-authoring, akin to a diary or narrative journal of the life of their map(s). Through interviews and participant observation, Del Casino and Hanna (2006) used performative and ethnographic methods to explore their 'map spaces'. Chris Perkins writes too that performative approaches may be fruitful as these 'see mapping as not only taking place in time and space, but also capable of constituting both' (Perkins, 2009: 2).

In a recent paper, Bittner, Glasze and Turk (2013) discussed the applicability of Laclau and Mouffe's theory of discourse and hegemony as a way of conceptualising the contingency of mapping projects. This takes the idea that the world is constituted by linguistic and extra linguistic articulations that become fixed at moments of 'sedimentation', but that these discourses are always able to be contested. As a consequence, discourses, in this case maps or elements within maps and the map making process, are contingent upon assemblages of actors and practices. Again, in order to understand what the map is we need to analyse not only the map but the suite of people and practices that articulate and contest it.

Bittner, Glasze and Turk (2013) engage actor-network theory as a means of gathering together the actors and forces that influence a map, and hence conceptualise a map assemblage. ANT has been applied to examinations of contemporary cartography by several authors including: Chris Perkins in his study of mapping golf (Perkins, 2006) and Francis Harvey's analysis of the GIS in administrative contexts in the USA (Harvey, Kwan and Pavlovskaya, 2005). In their 'Manifesto for map studies', Dodge, Perkins and Kitchin (2009) suggest that network analyses could be particularly appropriate for studies of collaborative and open-source projects. For the study of digital spatial media, and more specifically interactive and collaborative mapping projects, ANT offers a useful means of tracking and linking the various actors, including non-human actors (like datasets or software programs) as they affect a map over time. The network can encompass map producers, map users and the wider assemblages of actors supporting map production and use. It is a good way of sketching out myriad data sources and thinking about the chains of technologies that lead to a final map image.

Yet there are also limitations in using ANT to examine map assemblages. Foremost, while ANT helps us to consider and account for a broad range of actors, it does not, to my mind, provide a nuanced way of considering the relationships between actors. Despite attempts to differentiate and visualise the connections between actors, like that in Figure 9.3 from Jean-Christophe Plantin (2012), it is difficult to incorporate *how* organisations and things are linked to each other.

Moreover (and critically in the context of this chapter and book) it is difficult to incorporate *changes* in these relations, and hence to demonstrate how different actors work to produce maps or how these roles evolve *over time*. Returning to the question of how we account for the mutability of maps, we need rather a methodology which helps us to incorporate ideas of dynamic interfaces as well as shifting constellations of actors. Because maps are fluid, ontogenetic and

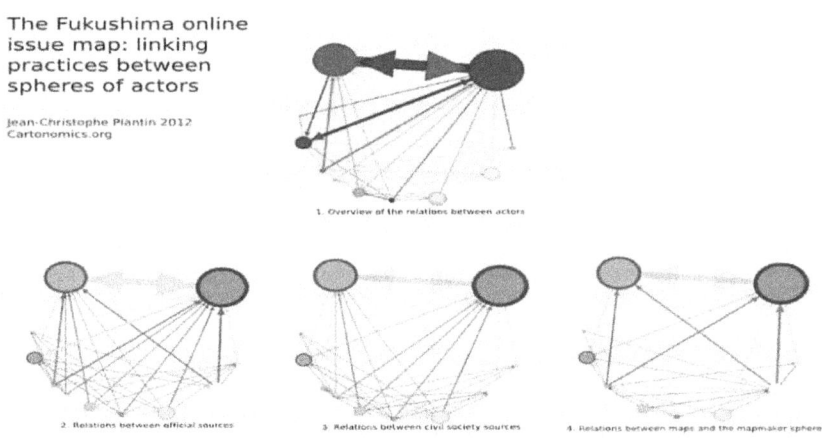

**Figure 9.3** Differentiating relations in a network (from J. C. Plantin, 'The Fukushima online issues map: linking practices between spheres of actors', *Cartonomics: Space, Web and Society*, 2012).

contingent upon networks of relations, such a methodology should help us to identify factors that affect a mapping project through time, such as what leads to stability, rupture, wide acceptance or use.

## Bubbles and foams

Philosopher Peter Sloterdijk's concepts of 'bubbles', 'spheres' and 'foams' offer just this: a way to conceptualise contingency and temporal variability in maps. In three connected books (Sloterdijk, 1998; 1999; 2004), Sloterdijk uses three sorts of sphere metaphors, or 'thought figures' as he prefers to call them (Funcke and Sloterdijk, 2005: 4), to examine social relations. The 'bubble' stands for the internalised world of the individual, insulated within a membrane yet in dyadic relations with the outside. 'Spheres' describes the idealised spaces of modernity, as all-encompassing universes; and 'foams' poses an alternative geometry for social relations where individual (bubbles) jostle within a 'multi-chambered system' (Sloterdijk, 2009: no pagination). Sloterdijk states:

> I try to describe these multiplicities of modern life in terms of foam-making – all individuals are living in a specific bubble within a communicating foam. (Sloterdijk, quoted in Morse, 2009: no pagination)

This idea of individual 'bubbles' within collective 'foams' has been taken up within cultural and architectural theory. Hélène Frichot (2009: 4) writes: 'contemporary society in the habitat of the city can be really said to behave in this way, like seething foam, co-isolated bubbles networked in ... clusters and symbioses'. While my discussion of maps and mapping practices does not describe society at large and operates on quite another level, I find the concept compelling and applicable to the study of contemporary cartography, precisely because it is able to accommodate a communicating assemblage that 'seethes' through time. Mapping projects, like 'bubbles', depend upon internal substance as well as relationships with outside networks or 'foams' of actors. We can think too about the 'bubble wall' as an interface between the map and these actors, and ask questions about the quality of the interface, its stability and continuing utility through time. Third, a 'foam' is fluid and dynamic, seething as 'bubbles' merge and expand, fluid interfaces recombining to make new 'bubbles'. There is the interplay of surface tension in bubbles and jostling from neighbours, sometimes shaped by larger forces but also capable of maintaining a stable form for a while. In the next sections, I look at these three aspects of the metaphor (networks, interfaces and dynamics) in turn, before coming back to the idea of temporality and going on to apply the maps as foams metaphor to a case study.

## Networks, assemblages, foams

Taking a map as a 'bubble' in an actor-network of 'foam', I imagine individual maps as bounded objects with particular combinations of content, sitting within networks of other bubbles. As Frichot (2009: 3) suggests: 'the operational analogy of foam offers an alternative model to help understand these networks of humans and things'. Applying the foam metaphor, we can see how a map is not only connected to other actors, but created through its relations with them. A map is a contingent object held together by data, technology, use and reputation. It is formed through the interfaces with other actors. It is as if we inflate the nodes in a network such that rather than connections being the thin strings of a web, each entity shares an interface with the neighbouring actors. Figure 9.4 shows the difference between an illustration of the network of actors combining in a mapping project (taken from Bittner, Glasze and Turk, 2013) to a foam of the same actors, as a basic visualisation of the shift in conceptualising this assemblage.

This step is a critical difference from the network in that the very objects are constituted by their relations with other things. I find the emphasis on contingency useful in directing my research, not just to gather together all the actors involved in a mapping project, but to critically examine the ways in which they relate to each other. The surfaces connecting bubbles are interfaces that can be observed over time to gauge how other actors (other bubbles) affect the shape of the bubble, its stability and longevity – and ultimately its existence.

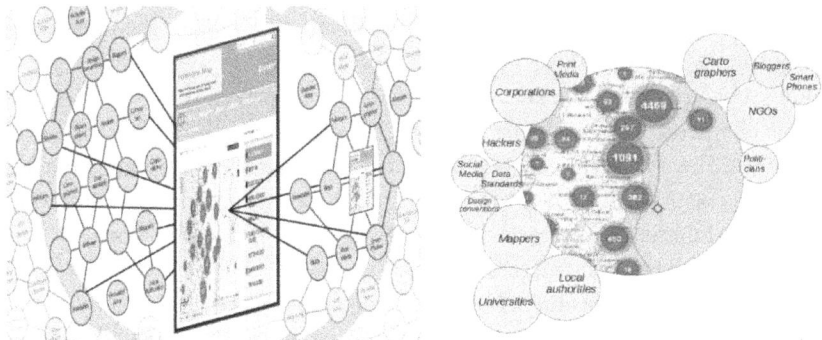

**Figure 9.4** Assemblages – from network to foam. Based on original map by S. Adler, G. Glasze, C. Bittner and C. Turk.

## Interfaces

Sloterdijk cites the architect Le Corbusier's reference to a soap bubble and his remark that: 'the outside is the product of an inside' (Sloterdijk, 2009: no pagination). This is true of a single bubble, where internal dynamics affect ultimate stability. But it is different in a foam: the outside of the bubble connects it to other actors. Bubble walls are interfaces, surfaces of exchange, representing continual to-ing and fro-ing between actors. These surfaces are the condition of contingency, as the means by which external actors affect the shape of the map-object. They are also a lens or film that mediates interaction. From the perspective of being within the foam, looking through the film from one bubble into another, the interface determines how neighbours exchange or view each other's content.

Thus, there are two ways in which we can make use of the idea of interfaces when imagining maps as bubbles – first, by examining the quality of relations between organisations and people and the exchanges or flows between them; and second, thinking more specifically about mapping, we can examine quite literally the quality of the mapping interface, such as cartographic design and usability. In so doing we need to take into account the multiple perspectives of users, imagining multiple positions from which we might look through interfaces into other bubbles. This ability to take positions is important to the study of interfaces, as Dodge and co-authors point out: 'interfaces en-frame and exclude, working as mediating windows onto the world ... [thus] ... the task of decoding the embedded cultural biases and distortions in processes of interface screening is challenging' (Dodge, Perkins and Kitchin, 2009: 222). Conceptualising interfaces as constituted by the interactions between actors means we are able to incorporate multiple (and hence biased) viewpoints.

Map interfaces are sites of exchange where information or influence moves between one entity and another. Rather than a strict boundary, the interface is a permeable structure made up of the exchanges to and fro between actors. Science studies scholar Andrew Pickering (2005: 21) has described these exchanges as a 'dance of agency' and, adding a temporal dimension, he shows how this dance constitutes the ongoing process of practice. Through working things out with others, an entity changes over time. Thus, the acknowledgement of contingency contains an inherently temporal dimension. Conversely, as Clive Barnett (2004: 17) describes, 'the poststructuralist understanding of temporality is in terms of a series of successive moments of pure contingency, tied together by nothing other than the force of an imposed convention or act of vitalistic will'. Coming back to our 'thought figure': through processes of articulation, the bubbles take on shapes and positions within a foam, yet these are, as Barnett (2004: 8) writes, 'only ever according to a contingent set of identifications that remain open to contestation'. The foam demonstrates the temporary nature of (map) assemblages, particularly the fluidity of interfaces/relations between actors.

## Dynamics and temporality

> Time emerges from a process that flows through the nexus of perpetually perishing (and perpetually becoming) actual entities. (Barker, 2012: 1051)

Studying interfaces as constituted by the relations between actors in a foam leads then inevitably to a consideration of dynamics and temporality. The constellation of bubbles and the quality of interfaces are continually shifting, such that the contingent assemblage moves like a fizzy foam. There are a variety of ways in which we might observe how mapping projects (as foams) undertake transitions temporally. The bubble metaphor invites consideration of surface tension, playing off interior and exterior pressures. We might also study how map bubbles 'get oxygen' so to speak, or inflate through self-promotional 'hot air'. Even the stable maintenance of relations is a continual dance of agency between actors through time. Foam physicists (*rheologists*) describe transitions where bubbles switch neighbours or where bubbles are subsumed into others. Some of these may be appropriate descriptions of shifts between actors affording, building upon or diverging from each other in the course of a mapping project. Maps gain authority and stability from the stability and quantity of users or supporters, this too changes over time and affects the form of the map.

There are thus a variety of shifts or transitions we could use to describe the evolution of a mapping project. The complexity of relations is, however, only able to be captured approximately in the metaphor of the mass of foam. The messy frothing foam signals that multifarious processes take place simultaneously, some connected, others with slower knock-on effects. A research methodology might focus in on particular sets of exchanges and consider how they are part of interrelated dynamic processes, happening at different speeds and scales. Tim Barker describes the multiple temporalities of digital media, where 'the time of the software, the time of the network, and the time of other users are all put in relation and are experienced as a mesh of multiple domains of the temporal' (Barker, 2012: 1586). Examining these, we can also draw on James Williams' discussion of Deleuze that introduces 'a formal network of processes defining time as multiple ... [operating] on one another but in different ways depending on which process takes another within itself' (Williams, 2011: 3).

Deleuze provides three different 'syntheses' of time that unsettle the concepts of 'past', 'present' and 'future' and examines how these concepts act upon each other as dimensions synthesised through particular processes (Williams, 2011). For example, Tim Barker suggests that: 'the digital medium is not merely a means of representing the world. Rather it is a mode of recomposing the present, of providing a means to rethink the present' (Barker, 2012: 102–103). He goes further to say digital media 'provides the potential for the actualisation of events and a state of presentness in which the past is constantly re-presented in the present' (Barker, 2012: 74). The endeavour to provide maps in 'real-time' underscores the emphasis on *representing* in online interactions, as previous datasets, programmed code and even future scenarios are drawn into suites of information exchanged in a present engagement between user and digital interface. In Barker's terms: 'the human user is temporalised by the digital process as his or her actions become transposed into the digital and this action alters his or her movements in the present' (Barker, 2012: 62).

Within the 'foam', we could account for multiple perspectives to examine how different actors in the assemblage 'temporalise' each other. How do past mappings work on new data? How do programming protocols produce particular ways of moving through sequences of dynamic maps? How are map curators temporalised by the flows of new map data? How does the presumed 'presentness' of real-time mapping feed into map use? Furthermore, acknowledging the multiple temporalities of digital spatial media, we might ask how these interactions are happening at different scales: from the micropolitics of interactions along 'bubble' interfaces to the larger ebbs and flows of the collective. Yet it is easy to get carried away with metaphors. The best test of utility is to see how the concept works with a concrete example, whether the 'thought figure' helps in shaping a research method.

## Crisis mapping foams

Natural catastrophes, or crises, like earthquakes, typhoons, tsunamis and bushfires, are heightened events where time is said to be 'of the essence' in coordinating a response to save lives and property. Here I examine how dynamic maps, digital spatial media, are being used to respond to such crises. I use my approach to the analysis of interactive 'crisis mapping' projects as a means to explore and review how using the concepts of 'bubbles' and 'foams' can help us to make sense of these mapping projects over time. The mapping of crises is an apt case study because we see how maps seek to account for shifting landscapes and changing circumstances, the temporal emergence of a crisis is echoed in the ways mappings emerge during crisis response.

Maps are an essential medium for organising and sharing information in emergency contexts – think of the big wall maps common in emergency coordination centres. *Crisis maps* are online collaborations where volunteers create maps to help understand and respond to natural disasters and political conflicts. For example, following the huge storm Typhoon Haiyan (locally known as Yolanda) which hit the Philippines in November 2013, several maps were created to help make sense of the crisis and to coordinate the aid response. Immediately after the storm, volunteer cartographers travelled to the Philippines to supply maps to the emergency responders. At the same time, mappers both amateur and professional worked together online to map the catastrophe from afar.

Here, through a 'foam' inspired investigation, I track these projects, exploring the actors they bring together, the way maps gain traction among actors (and hence relevance to the disaster response) and how this changes over time. Following archival research and participant observation, a range of mapping projects have been analysed (see Table 9.1 below).

Employing a methodology inspired by the bubble/foam metaphor, for each of these examples might usefully deploy questions documented in Table 9.2.

**Table 9.1** Archival mapping sources used by author for analysis

| Mapping project | Source |
| --- | --- |
| Humanitarian OpenStreetMap Team (2013) | https://wiki.openstreetmap.org/wiki/Typhoon_Haiyan |
| VISOV (Volontaires internationaux en soutien aux opérations virtuelles)/ Ushahidi Crowdmap (2013) | https://haiyan.crowdmap.com |
| University of Heidelberg/Ushahidi (2013) | http://crisismap.geog.uni-heidelberg.de/ushahidi/login |
| StandbyTaskforce/GIS Corps/ESRI (2013) | www.esri.com/services/disaster-response/hurricanes/typhoon-hayian-yolanda-maps |
| Google Crisis Response (2013) | http://google.org/crisismap/a/gmail.com/TyphoonYolanda |
| Tomnod Satellite Image assessment (2013) | www.tomnod.com/campaign/haiyantyphoon2013 |
| Philflood map/Ushahidi Crowdmap (2013) | https://philfloodmap.crowdmap.com |

**Table 9.2** Questioning the mapping bubbles and form

| | |
|---|---|
| *Internal substance:* | What is the content of the map? |
| | What form does the project take? |
| *Relationships between networks/foams of actors:* | Who is involved in making the map? |
| | Who uses it? |
| | How established are its supporters? |
| | Are potential actors left out? |
| | How does this change throughout the project? |
| | What sorts of practices maintain the foam? |
| *The quality of the interface(s):* | How is the map used? |
| | How is it communicated? |
| | How stable or credible does it seem (to other actors)? |
| | How do the actors interact with the map and with each other? |
| *Continuing utility through time:* | How has the map changed? |
| | How has the 'foam' around it shifted? |
| | What are the common understandings that hold a foam together? |
| | What sorts of data flows have been incorporated? |
| | What sorts of feedback processes does this entail? |

Through this sort of analysis, it is possible to draw out factors that have influenced the ability of these crisis maps to reach their set objectives, considering, at the same time, barriers and rupture. Some of this research is presented here, demonstrating how the 'foam' metaphor and the emphasis on networks, interfaces and dynamics might be applied to an analysis of mapping.

## Networks

Probably the most highly publicised map response to Typhoon Yolanda/Haiyan was the work of Humanitarian OpenStreetMap Team – HOT (2013). The 'Hotties', as they are known, engaged volunteers who worked together online to help trace map features from satellite imagery to create maps of the affected areas (examples of the HOT maps are shown below in figures 9.5, 9.6 and 9.7). This sub-group of the OSM community has established protocols for responding to crises and is part of international networks of institutions assisting disaster response. Furthermore, it has an established base of volunteers, who have mapping and programming skills, as well as resources for quickly training newcomers to help map. Communication channels are well set up and those contributing to the map can make use of an interface that has been evolving through user feedback. As a result, the HOT team was able to rapidly supplement existing maps of the affected areas (using pre-disaster satellite images from Bing maps) and then, in a second phase, map the extent of damage because they were granted access to post-disaster satellite imagery.

This mapping project 'bubble' is well supported within a relatively stable 'foam' of significant long-term actors and many smaller contributors. The reputation of the existing organisation, the number of people contributing and the stability of key actors supporting this mapping project are all factors that have promoted the project's longevity and success, not only in getting the mapping work done, but in sharing the map with users (both in the disaster affected areas and elsewhere). The HOT map stands in contrast with maps that have had poor uptake due to a lack of engagement, such as https://philfloodmap.crowdmap.com/ where there have been few entries and hence a very small network of adjoining bubbles.

## Interfaces 1

Often the quality of the interface plays a role in determining a map's wider circulation and use. This Google map (http://google.org/crisismap/a/gmail.com/TyphoonYolanda) has large icons which often overlap, making it hard to recognise resources available or get an overview of the situation. Thus, despite the overwhelming reputation of the parent organisation, the map has limited use. Another map drew on a popular crisis mapping software (Ushahidi), and was instigated by a reputable academic institution (University of Heidelberg, 2013) – both established actors – but password access meant that potential users were vetted. The audience of passive map viewers (who might have browsed the map but didn't want to register) was thus also restricted, limiting the circulation of the map. In these last two cases the interface – the ability of others to interact with the map – determines who is included in the (foam) assemblage and the terms of the relations. It is worth noting these are reciprocal conditions: the more actors there are, the more information that is able to be contributed to the map, and therefore it is more likely to attract further users. We might think of this as the 'stickiness' of the bubble assemblage.

## Interfaces 2

More is revealed about these projects by examining the other sense of 'interfacing'; how map users or contributors interact with each other. Rob Kitchin and Martin Dodge have noted the way in which crowd-sourced projects, like OSM, demonstrate shifting ontologies of mapping projects because they expose the decision-making processes and contingencies in map production. They write: 'OpenStreetMap is a valuable "live laboratory" in which to explore the ontological politics of cartography and the ways in which a mapping ontology "appears solid, but is in fact always shifting"' (Dodge and Kitchin, 2013: 29). Because discussions about decisions are archived in emails lists and a wiki, we can see through the discourse how and why curatorial choices were made. Actors work contingently to affect the OSM map making process, jostling in a way we might imagine as akin to bubbles shaping each other in a foam.

In this case, a chain of emails between participants who helped HOT's online response to Typhoon Haiyan/Yolanda reveal some of the micro-dynamics of this mapping project: from the initial email before the storm warning alerting members of the possibility that their help may be needed, to the negotiations of specific mapping practices, such as how to label damaged buildings or gain access to satellite imagery. Emails written by different users give us positioned descriptions of the process, thus we can see how those new to HOT or OSM sought clarifications about mapping protocols, or those representing aid organisations put questions to the HOT community. Feedback from NGOs using the HOT map in the affected area helped to inform the way map features are edited, and some of the online mapping community even found opportunities to undertake fieldwork, further enriching the map. These processes of feedback at the interfaces (exposed for my research in this case in the HOT email list, but taking place continuously through discussions and interactions between actors) describe the continual shaping of the map.

## Dynamics

Taking these exchanges as a starting point, the following sections and the accompanying cartoons (see Figures 9.5, 9.6 and 9.7) focus on micropractices at the interfaces between actors. They serve to demonstrate ways in which this approach helps to study the dynamics of mapping projects. Three different perspectives are presented examining: dynamics between actors; dynamics within a map; and dynamics of 'foams', respectively. Each shows how the 'foam' metaphor enables a consideration of temporality and changes in relations.

### 1 Dynamics between actors: shaping networks over time

The HOT mapping project, as described above, is an example of a well-supported and used map. The stability and established reputation of organisations associated with HOT has helped to circulate the map among emergency responders working on the ground in the Philippines, as well as among those

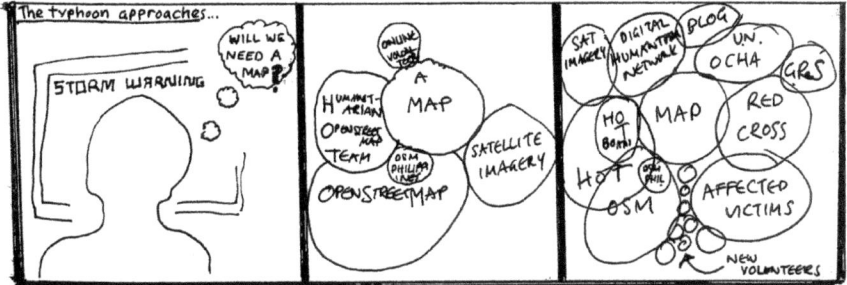

**Figure 9.5** The HOT mapping deployment post Typhoon Haiyan/Yolanda: inception and growth (author's image).

coordinating the response. Here, the foam illustrates an assemblage where the support of significant actors gives credibility and extends the possibilities for wider map distribution and engagement. These stable actors give support to the project through the stability of the interface.

Yet there may also be dynamics that lead to the possibility of rupture. At one stage, the HOT email list shows frustrations felt in waiting for access to satellite imagery. While some of the coordinating group were undertaking negotiations with larger institutional partners, others set up a petition to try and lobby for speedier access to the imagery. Emails allude to tensions between the HOT board and a petitioner. In this case the bubble doesn't split but stability is disrupted.

Another example shows dynamics between actors and how we can take positioned viewpoints into account, looking through the foam.

This sequence of events shows how the network of significant actors, while promising stability, poses some risks for how the HOT map is viewed. In this case, a partnership within a network of emergency response organisations

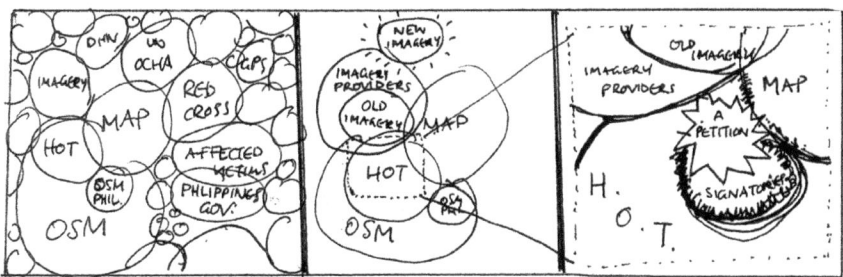

**Figure 9.6** The HOT mapping deployment post Typhoon Haiyan/Yolanda: discussion and disruption (author's image).

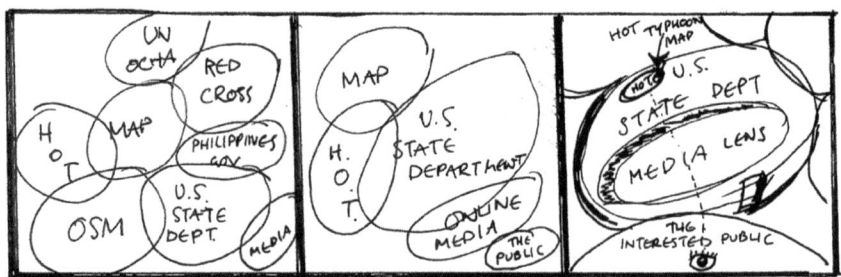

**Figure 9.7** The HOT mapping deployment post Typhoon Haiyan/Yolanda: viewed through other actors (author's image).

and, in particular, a close engagement with the United States government's State Department, potentially overshadows the work of HOT. The ambiguous phrasing of this article (DipNote, 2015) about HOT work, in the context of a joint project with the State Department known as 'Mapgive', implicitly credits the US Department of State for the HOT's Haiyan/Yolanda response (Amrwaga, 2015). While the State Department would most probably clarify the nature of their engagement, from the perspective of an outside actor reading about these mapping activities, the State Department 'bubble' takes in the HOT work. These small interactions, insignificant in the overall work that these maps set out to do, are relevant here because they signal how a map is contingent upon the ways it is generated, presented and used.

## 2 Time within a map

Maps are opportunistic compilations of data, 'bricolage' as John Pickles puts it (2004: 88; following Lestringant, 1994). As such, data from different sources and time periods may be combined together into a map, sometimes in a chronological sequence, but often containing disjunctures and inconsistencies. Take as an example this screenshot (Figure 9.8) from the Volontaires internationaux en soutien aux opérations virtuelles (VISOV, 2013) Ushahidi Crowdmap of damage from the typhoon.

The screenshot shows map data in the form of a list of reported entries, mainly incorporating photographs of damage. Reports are date stamped at the right according to when they were entered into the 'crowdmap' database. Some reports contain dates as part of the data entered – see 'Nov 11 AFP' in the third entry, for example. We could assume this was when the photo was taken, but it could also be when the organisation received the photographs from someone else. So there are least two different times combined into the photo report and

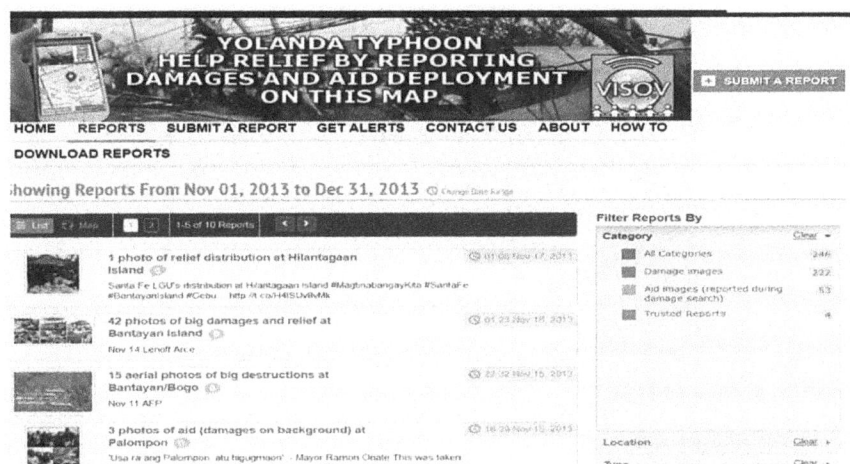

**Figure 9.8** Mapping software collating crowd-sourced reports about storm damage (Volontaires internationaux en soutien aux opérations virtuelles, https://haiyan.crowdmap.com).

going back in the chronology there is also the time when the damage occurred. There will always be a time lag (no matter how sophisticated the software) between the moment an event happens and when it is recorded. In the online context, there is also the further issue of how the software copes with multiple time zones and correspondences between the time of the server and of the user. This series of times and events then becomes collapsed into a report which, in turn is translated into a dot on the map.

The crowdmap software enables users to filter reports according to date – noted here by the red text 'Showing Reports from Nov 01, 2013 to Dec 31, 2013' – such that a map image might only show a subset of reports. The map database, however, is an accumulation of reports over time (see Figure 9.9). Curators may have the ability to delete reports, but in practice most crowdmaps consist of a piling up of data, such that calling up a map in the present includes old and new data, some potentially out of date. There is therefore a need to interpret maps critically and ask whether the current version reflects a real-time present.

The processes of data capture and representation draw together events happening at different times and reconfigures them in a map interface. Again, employing the metaphor of a foam assemblage suggests honing in on particular practices that contribute to the temporality of a map through these technological engagements.

Like the adding of damage reports through the software interface, a second example of complex temporalities feeding into the map is that of mapping

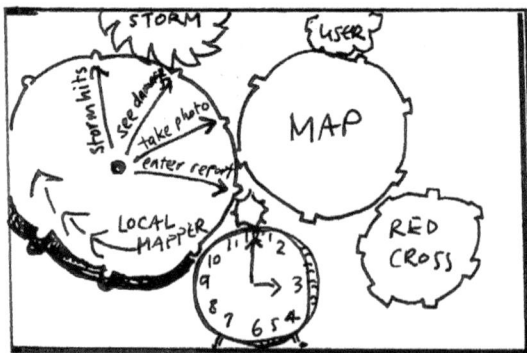

**Figure 9.9** Foam interfaces can be thick with multiple temporalities (author's image).

damage from satellite imagery. A couple of the case study map examples make use of satellite imagery of the damaged areas as it becomes available. The Tomnod project (Digital Globe, 2013), one of these examples, invites members of the public to scan satellite imagery, crowd-sourcing the job of processing hundreds of square kilometres of land to recognise damage from the air. Taking a performative approach to these practices of doing the scanning work, we see how this process of map or image interpretation involves toggling backwards and forwards between pre- and post-disaster imagery. By undertaking this jumping through time, the user sees and experiences, albeit through a mouseclick, the effects of the storm. Each new tile viewed begins this process of time travel and (armchair) experience of the disaster again.

The HOT email list shows there was a considerable drive to achieve up-to-date maps using the latest satellite imagery (and this in a timely way), but even in this striving for proximity to real-time, the present is mapped with reference to the past. In order to map what is damaged we need to know what was there before. It is important to remember this, given the internet's sense of always being in the present, meaning what we access today we often assume to be current. This is particularly pertinent in the context of a 'crisis' where the urgency of response is a plaidoyer for attempting to map in real-time. Yes, in comparison to a static wall map, digital spatial media offers possibilities for integrating new data as it is produced, including incorporating non-human sensors. Further, the ability to animate the map provides an interface that seemingly moves with the times. We imagine real-time as represented by the moment new data pops into a map on our screen. Still, David Pinder's warning (2007: 460) not to get caught up 'in the giddy spectacularised techno-rush that promises even more powerful techniques of visualisation', is sound advice. In a way, as we focus in on the micropractices of mapping and take a 'freeze frame' in order to interpret the gushing, foaming streams of crisis data, we heed his call for slowing down and pausing.

## 3 Foams over time

Yet how might we take an assemblage of seething foam into account? One way is to recognise and chart the multiple temporalities working on each other, such as the ways in which older data is incorporated, or how remarks about the present rely on knowledge about the past. As a crisis develops and mapping effort follows as part of the response, the expected temporality of rapid exchanges and updates may be frustrated or disrupted; the tempo changing depending on accessibility of information about the crisis or resources to help contribute to a map. Temporality is thus a significant dimension in the relations between actors in crisis mapping assemblages.

It is also possible to conceive of the temporality of the foam at another scale, not just in exchanges between individual actors, but as a way of considering how the collective moves and changes. From the jostling of bubbles and study of surface tensions, we could, maintaining the metaphor, go on to examine the larger 'oceanic' movements; the forces that churn the foam and affect the longer term progress of these projects (see Figure 9.10). Oceanic metaphors have also been used to refer to big data – we hear for instance about a 'tsunami of information' (D'Antonio, 2011; Brown, Vinzi and Glady, 2013). Crisis map project-histories could very well be put in similar terms coming into being on a tide of concern; being brought to the fore through media interest in the disaster, churning and collecting more actors; and sitting in clusters that after time and exposure become subsumed into the ether.

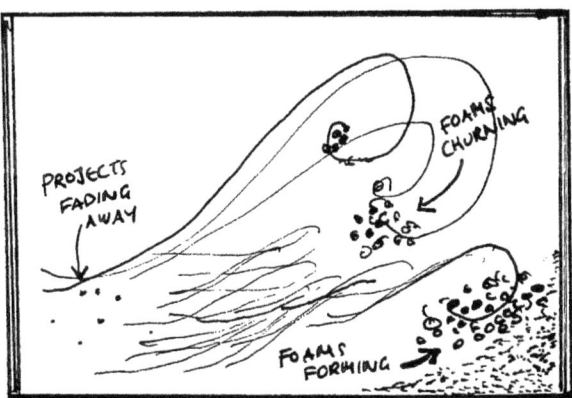

**Figure 9.10** Map project histories, with apologies to Hokusai (author's image).

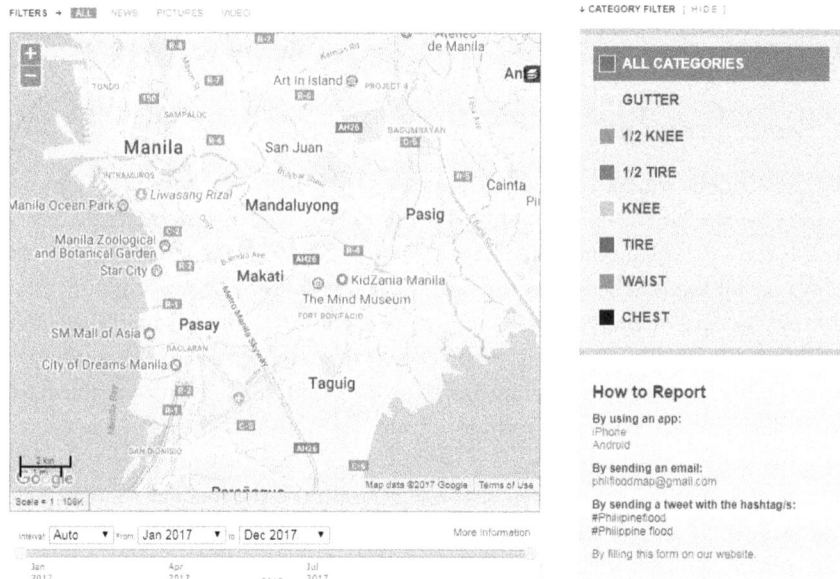

**Figure 9.11** Crowd-sourced reporting of water heights. Information is current for a limited time. Philippine Flood Map 2013, https://philfloodmap.crowdmap.com. Map data © 2017 Google. This figure has not been made available under a CC licence. Permission to reproduce it must be sought from the copyright holder.

There is too the ephemeral nature of these maps and their overall relevance in the big scheme of disaster response. These maps are often a 'drop in the ocean', they have a moment of exposure and fizzle away. Unless a significant actor or a great many users latch onto the map, it is taken back into the sea of data. In the life of a 'fizzing map foam', new actors join, others leave, engaging or disengaging with the interface. Some maps, like the Philfloodmap example in Figure 9.11, are designed for a short period of use. In this case users should report in when water reaches a certain height.

The information is relevant for a brief, critical period and depends upon user input during this time in order to work. If there is poor exposure or users engage only briefly without contributing reports, then there is little content. The map may be subsumed among other sorts of crisis information and fizzle away.

Here the emphasis on the dynamics of map project history provides a more nuanced account of these maps and the work of those creating and employing them. As a method for studying maps, we can include a level of detail accounting for the multiple temporalities of map assemblages at different scales. This is particularly relevant to crisis mapping; where maps can be generated and edited quickly and the situations in which maps are used may also be subject to rapid change. Often maps are being made in two time frames at once – sourcing pre-disaster information and attaching post-disaster reports. Furthermore, the constellations of actors move and slip as aid agencies, local communities and online collaborators join, make use of, or discard the maps produced. The foam provides a figure that helps to encompass these actors and actions, while maintaining a central interest in the map.

## Conceptualising maps that shift with each moment

> Once you begin a hunt for spheres in the form of bubbles, globes and foam, and so forth, they seem to spring up everywhere, appearing in all variety of shapes and sizes and inaugurating all kinds of relations. (Frichot, 2009: 4)

Sloterdijk's 'thought figure' is compelling. The bubble/foam topology is an attractive way to conceptualise an assemblage. As Sloterdijk (2009: no pagination) himself has remarked however: 'it is not a universal theory but an explicit form of spatial interpretation'. Having explored the possibilities here, I consider it offers a useful way to organise research about relations between objects and the networks/foams within which they exist. Most importantly for our discussion here, the foam metaphor helps to convey the contingency of mapping projects and the dynamism of relations. Not only the dynamics of map content within new spatial media, but significantly the dynamism inherent in all maps. Furthermore, we can incorporate concepts of interactivity; bricolage (reuse and recombination of data etc.); stability and fragility; manifold perspectives and the mediation of interfaces; as well as ideas of (surface)tension and exchanges of agency. Not to mention the possibilities of viscosity: how foams fizzle and seethe according to multiple temporalities.

Through the example of mapping Typhoon Haiyan/Yolanda, we have been able to showcase some of the methods by which the foam metaphor can be applied to help shape the inquiry and better understand the dynamic relations within crisis mapping projects. This has helped to show projects where there is stability and good reception of the map, as well as how maps have utility for short periods of time or specific groups of users. As Annemarie Mol and Marianne de Laet say of the fluid networks surrounding the technology of the Zimbabwe bush pump (Mol and de Laet, 2000), there is no binary assessment of whether these maps are successful or not. Rather the researcher allows herself/ himself to be moved by them. Along these lines, I should acknowledge that the fluid mass of foam includes me as researcher, my views on how these crisis mapping projects look and work are positioned within the foam too.

Considering digital spatial media as combining multiple temporalities, it is worth noting how data and relations accumulate in crisis maps over time, as well as how collaborating in real-time intersects with the contingency of mapping projects. Employing a method inspired by foams, we have accounted for these dynamics at different scales: both the micropolitics of interactions at the interface and the larger ebbs and flows of the collective. The transition from 'networks' to 'foam' incorporates concepts of 'ontogeny', and enables the processual approach to understanding maps to account for project histories. The jostling of bubbles in the foam is a potent way of imagining maps and engagements with maps, including the flurry to map order into a crisis.

## Note

1  The term 'produsers' is used by Bruns (2006: 276) to reference a hybrid of map producers and users.

## References

Amrwaga (2015) Open data day: How the State Department is linking diplomacy with collaborative mapping during crises. *Str8talk Chronicles*, 23 February. [Online] Available at: http://str8talkchronicles.com/open-data-day-how-the-state-department-is-linking-diplomacy-with-collaborative-mapping-during-crises (accessed 1 March 2015).

Barker, T. (2012) *Time and the Digital: Connecting Technology, Aesthetics, and a Process Philosophy of Time*. Hanover, New Hampshire: Dartmouth College Press.

Barnett, C. (2004) Deconstructing radical democracy: Articulation, representation and being-with-others. *Political Geography*, 23(5): pp. 503–528.

Bittner, C., Glasze, G. and Turk, C. (2013) Tracing contingencies: Analyzing the political in assemblages of web 2.0 cartographies. *GeoJournal*, 78: pp. 935.

Brown, S., Vinzi, V. E. and Glady, N. (2013) Big data: A tsunami of information. Essec Business School. [Online] Available at: http://knowledge.essec.edu/en/innovation/keeping-up-with-the-big-data-revolution.html (accessed 1 March 2015).

Bruns, A. (2006) 'Towards produsage: Futures for user-led content production'. In: Sudweeks, F., Hrachovec, H. and Ess, C. (eds) *Proceedings: Cultural Attitudes towards Communication and Technology*. Perth: Murdoch University, pp. 275–284.

Cartwright, W. (2013) 'Artefacts and geospaces'. In: Kriz, K., Cartwright, W. and Kinberger, M. (eds) *Understanding Different Geographies: Lecture Notes in Geoinformation and Cartography*. Berlin: Springer-Verlag, pp. 55–66.

D'Antonio, M. (2011) Marketers brace for the Tsunami of big data. *1to1Media*. [Online] Available at: www.1to1media.com/view.aspx?docid=33289 (accessed 1 March 2015).

De Laet, M. and Mol, A. (2000) The Zimbabwe bush pump: Mechanics of a fluid technology. *Social Studies of Science*, 30(5): pp. 225–263.

Del Casino, V. and Hanna, S. P. (2006) Beyond the 'binaries': A methodological intervention for interrogating maps as representational practices. *ACME: An International E-Journal for Critical Geographies*, 4(1): pp. 34–56.

Digital Globe (2013) Tomnod satellite image assessment. [Online] Available at: www.tomnod.com/campaign/haiyantyphoon2013 (accessed 1 March 2015).

DipNote (2015) Open Data Day: How the State Department is linking diplomacy with collaborative mapping during crises. U.S. Department of State Official Blog. [Online] Available at: http://2007-2017-blogs.state.gov/stories/2015/02/21/open-data-day-how-state-department-linking-diplomacy-collaborative-mapping-during.html (accessed 5 December 2017).

Dodge, M. and Kitchin, R. (2007) Rethinking maps. *Progress in Human Geography*, 31(3): pp. 331–344.

Dodge, M. and Kitchin, R. (2013) Crowdsourced cartography: Mapping experience and knowledge. *Environment and Planning A*, 45: pp. 19–36.

Dodge, M., and Perkins, C. and Kitchin, R. (2009) 'Mapping modes, methods and moments: A manifesto for map studies'. In: Dodge, M., Kitchin, R. and Perkins, C. (eds) *Rethinking Maps*. London: Routledge, pp. 311–341.

Elwood, S. and Leszczynski, A. (2012) New spatial media, new knowledge politics. *Transactions of the Institute of British Geographers*, 38: pp. 544–559.

ESRI, GIS Corps and StandbyTaskforce (2013) *Haiyan Yolanda Perspectives Map*. [Online] Available at: www.esri.com/services/disaster-response/hurricanes/typhoon-haiyan-yolanda-perspectives-map (accessed 1 November 2014).

Frichot, H. (2009) 'Foaming relations: The ethico-aesthetics of relationality'. In: Meade, T (ed.) Occupation: Negotiations with Constructed Space, Conference Proceedings, University of Brighton, 2–4 July.

Funcke, B. and Sloterdijk, P. (2005) 'Against Gravity': Bettina Funcke talks with Peter Sloterdijk. *BookForum*, February/March. [Online] Available at: www.bookforum.com/archive/feb05/funcke.html (accessed 1 November 2014).

Harvey, F., Kwan, M. P. and Pavlovskaya, M. (2005) Introduction: Critical GIS. *Cartographica*, 40(4): pp. 1–3.

Humanitarian OpenStreetMap Team (2013) *Typhoon Haiyan*. [Online] Available at: http://wiki.openstreetmap.org/wiki/Typhoon_Haiyan (accessed 1 November 2014).

Kitchin, R., Gleeson, J. and Dodge, M. (2013) Unfolding mapping practices: A new epistemology for cartography. *Transactions of the Institute of British Geographers,* 38: pp. 480–496.

Lammes, S. (2011) 'The map as playground: Location-based games as cartographical practices'. In: *Think, Design, Play: Proceedings of the Fifth International DIGRA Conference*, Utrecht, pp. 1–10.

Latour, B. (1990) 'Drawing things together'. In: Woolgar, S. and Lynch, M. (eds) *Representation in Scientific Practice*. Cambridge, Massachusetts: The Massachusetts Institute of Technology Press, pp. 19–69.

Lestringant, F. (1994) *Mapping the Renaissance World: The Geographical Imagination in the Age of Discovery*. Translated by D. Fausett. Berkeley: University of California Press.

Morse, E. (2009) Something in the air. *Frieze Magazine*, 127. [Online] Available at: www.frieze.com/issue/article/something_in_the_air/ (accessed 1 December 2014).

Perkins, C. (2006) Mapping golf: A contextual study. *Cartographic Journal*, 43(3): pp. 208–223.

Perkins, C. (2009) 'Playing with maps'. In: Dodge, M., Kitchin, R. and Perkins, C. (eds) *Rethinking Maps*. London: Routledge, pp. 167–188.

Philfloodmap (2013) *Philippine Flood Map 2013*. [Online] Available at: https://philfloodmap.crowdmap.com (accessed 1 November 2014).

Pickering, A. (2005) *The Mangle of Practice: Time, Agency and Science*. Chicago, Illinois: University of Chicago Press.

Pickles, J. (2004) *A History of Spaces: Cartographic Reason, Mapping, and the Geo-Coded World*. London: Routledge.

Pinder, D. (2007) Review essay: Cartographies unbound. *Cultural Geographies*, 14: pp. 453–462.

Plantin, J. C. (2012) The Fukushima online issues map: Linking practices between spheres of actors. *Cartonomics: Space, Web and Society*. Weblog, 2 March. [Online] Available at: http://cartonomics.org/2012/03/08/post-fukushima-radiation-debate-mapping-the-online-issue-network/ (accessed 01 December 2014).

Sloterdijk, P. (1998) *Sphären I – Blasen*. Frankfurt: Suhrkamp.

Sloterdijk, P. (1999) *Sphären II – Globen*. Frankfurt: Suhrkamp.

Sloterdijk, P. (2004) *Sphären III – Schäume*. Frankfurt: Suhrkamp.

Sloterdijk, P. (2009) Talking to myself about the poetics of space. *Harvard Design Magazine*, 30. [Online] Available at: www.harvarddesignmagazine.org/issues/30/talking-to-myself-about-the-poetics-of-space (accessed 1 December 2014).

University of Heidelberg, GIScience (2013) *Training and Test Crisis Map of GI Science Heidelberg*. [Online] Available at: http://crisismap.geog.uni-heidelberg.de/ushahidi/login (accessed 1 June 2014).

VISOV (Volontaires internationaux en soutien aux opérations virtuelles) (2013) *Yolanda Typhoon*. [Online] Available at: https://haiyan.crowdmap.com/ (accessed 1 November 2014).

Williams, J. (2011) *Gilles Deleuze's Philosophy of Time: A Critical Introduction and Guide*. Edinburgh: Edinburgh University Press.

Wood, D. and Fels, J. (2008) *The Natures of Maps: Cartographic Constructions of the Natural World*. Chicago, Illinois: University of Chicago Press.

# 10
## Maps as objects
*Tuur Driesser*

### The PathoMap

In September 2013, New York University (NYU) announced an award of $250,000 to two teams of researchers as part of the presidential Grand Challenge competition dedicated to promoting 'significant scientific research that has the potential to solve major national or global problems' (New York University, 2013: no pagination). One of these teams, comprised of researchers from a number of the university's research centres, including the Center for Genomics and Systems Biology and the Center for Urban Science and Progress (CUSP), won its award based on a proposal to analyse New York's metagenome in order to map the city's microbiome. This means that by analysing the genetic material found in environmental samples from subway stations and parks throughout the city and tagging these with GPS coordinates, a map is created indicating the locations of the city's microbial communities. With its capacity for data collection and analytics, CUSP contributes to Weill Cornell Medical College's 'PathoMap' project (Dale, 2013), which was an experiment that used next-generation DNA sequencing methods to eventually monitor pathogens and dangerous organisms in real-time as they emerge within a country or city. By the project's first evaluation (Afshinnekoo *et al.*, 2015), researchers had collected, catalogued and analysed a total of 1,457 environmental samples, providing a base line map of New York City's 'microscopic residents' (PathoMap, n.d.). In future stages, this project could enable first responders and specialised threat teams to act quickly to contain and remove a biological threat:

specifically, we propose to leverage our previous work in multi-layered microbe detection and human sequencing to establish a 'pathogen weather map,' or 'PathoMap,' whereby the microbiotic population and genetic dynamics in a city can be used to detect and respond to increases in microbial dangers. (PathoMap, n.d.)

Using the new possibilities offered by DNA sequencing technologies developed in the analysis of the human genome, in combination with GIS software, researchers hope to closely approximate a real-time map of New York's microbiological life. Ultimately, the PathoMap is aimed to function as a monitoring device to enable quick detection of pandemic threats – in particular in relation to bioterrorism – and subsequent intervention.

Here, this case of the PathoMap as a monitoring device will serve to explore some questions concerning maps and temporality. In particular, the argument will emphasise not the representation of time through maps, but rather the way in which maps themselves affect, direct and produce time. Promoting the Picturing Place project at the University College London 'Urban Laboratory', Campkin, Mogilevich and Ross (2014: no pagination) write that:

> [w]hen we make things visible, we make them public and subject to debate. Urban images not only offer insights to help analyse and understand cities, but also point to ways of radically altering their futures.

However, as will be suggested here, the way in which urban images – maps – can change these futures is not (only) by 'producing evidence for physical change' (Campkin, Mogilevich and Ross, 2014: no pagination) to make things public. Instead, more than an independent, transparent and accessible layer, maps take part in the production of urban space (increasingly so with the proliferation of digital visualisations). Thus, a map will be considered as part of an urban assemblage and, consequently, as a constitutive part of the city's functioning. That is not to say that the map and the city are the same, or exist only in relation to one another, but that their encounter is a specific constellation – sometimes temporary, sometimes more permanent. Within such constellations, through their visualisation mechanisms, maps draw together and drive apart; separate and augment (Hookway, 2014); enable and disable. In the case of the PathoMap, it will be explored how exactly the mapping of the NYC microbiome produces a particular configuration of space and time within this urban assemblage. The main issue thus addressed concerns the way in which the PathoMap establishes relationships between the present and the future of the city. The *critical* object-oriented question ultimately becomes how maps of the present can enable the opening-up or closing-down of the future and, consequentially, make possible (or impossible) a social imagination that takes seriously the openness of this future.

## From critical to object-oriented cartography

The critical cartography which arose in the 1990s (Crampton and Krygier, 2006) approach maps as texts (Harley, 1989), sign systems (Wood, 1993) and social constructions (Crampton, 2001). In response to the dominance of the communication model, which thought of maps purely as neutral tools to convey geographical information, critical cartography sought to demonstrate how these representations were in fact bound up with politics of power and knowledge. Thus, building on Foucault and Derrida (Harley, 1989), the project of critical cartography was to deconstruct the map in order to reveal it as a 'specific set of power-knowledge claims' (Crampton and Krygier, 2006: 12). Accordingly, within the field of critical cartography, research shifted from determining the most efficient methods for communication to examining how power finds its expression in maps. Yet despite their critique of representation, Kitchin and Dodge (2007) note that writers, such as Harley and even Crampton, still operate within an essentially representational framework. As such, one of the implications of emphasising 'intertextuality' for Harley (1989) was the possibility of looking beyond the map maker's intentions in order to find other, suppressed and competing, narratives. Here, this analogy of maps and texts by Harley will be revisited by examining maps through the lens of some recent ideas about literary criticism from 'new materialists', such as Bennet (2012), Harman (2012) and Hayles (2014). This will lead into a proposal for an object-oriented cartography in which maps again figure as texts, not with the purpose of deconstruction, but to argue against Wood's (1993: 81) assertion that 'their quality as things is less significant than their quality as signs'.

If the constructivist approach of critical cartography challenged the idea of neutrality of representation as formerly presupposed by the communication model, an object-oriented critical cartography contributes to destabilising the communicational function of a map. It describes how maps operate in all sorts of ways, besides merely conveying geographical information. Harman's (2012) object-oriented philosophy is useful here as it warrants the map against 'undermining' descriptions which explain objects in terms of their constituent elements, such as those attempts under the communication paradigm to improve map design by optimising its various components. At the same time, it warns against 'overmining' the map by explaining it as the obverse of some larger structures – e.g. as an expression of the mapmaker's culture, ideology, consciousness, and so forth. In opposition to constructivist frameworks, an object-oriented critical cartography affirms the map's connection to the territory. As part of the (speculative) realist project, it takes a step back from that all-encompassing prison-house of language to return some sense of ontological security to the territory. This is cartography as what Kurgan (2013: 34–36) calls a 'para-empirical' analysis: an 'effort at once to reclaim a sense of reality, and not to imagine that this requires doing away with representations, narratives, and images'. Acknowledging the inherently abstracting qualities of representation, it re-evaluates the relationship of the map to the territory as one that is a representation, but a representation that exists 'alongside the world' rather than on top of it.

Such a framework does not necessarily need to altogether reject discursive analyses in favour of a 'materialist' approach, but rather asserts the latter as a 'useful counter' to the former (Bennett, 2004: 358). The object-oriented approach, and the wider philosophical movement of speculative realism more generally, points towards the limits of deconstructive, language-focused critiques. With respect to maps, the deconstructionist work of the early critical cartographers has been fundamental for questioning the neutrality of maps. As Bryant (2013) describes, poststructuralist-oriented critiques of critical theory have been important in various emancipatory political struggles where it was necessary to demonstrate how categories such as gender and race were culturally constructed. Nevertheless, it is also clear that not all forms of power are necessarily reducible to issues of discourse. Harley (1989) briefly touches upon the 'power external to maps and mapping', referring to the deployment of maps for the exercise of power, mainly in the context of surveillance and control. It may even be granted that his distinction between internal and external power becomes rather blurred at times. The main point, however, is that while critical cartography is crucial to understand how power is expressed in seemingly neutral maps, its fundamentally constructivist focus falls short of addressing how the power of the map is exerted on the world in the first place.

Object-oriented cartography aims to confront this shortcoming of the limits of critical cartography's capability to explain the power of maps. Thus, moving beyond hermeneutic cartographic interrogations, maps can for instance be considered as: devices (Marres, 2012; Law and Ruppert, 2013), apparatuses (Agamben, 2009), objects (Harman, 2012) or things (Bennett, 2004). The function of these conceptualisations is to suggest that maps, possessing some degree of coherence, with more or less distinct boundaries, fulfil particular roles by their 'capacity to capture, orient, determine, intercept, model, control, or secure the gestures, behaviours, opinions, or discourses of living beings' (Agamben, 2009: 14) and, as should be added in this context, non-living objects. Most importantly, what these conceptualisations have in common is the emphasis on the need to take seriously the materiality of objects and their affective capacities. This framework is related to literature in post-representational cartography (Perkins, 2004), itself associated with the non-representational turn in geography, and the social sciences more generally. Rather than focusing on what is and what is not represented, it is key to understand 'how mappings emerge, circulate and do work in the world' (Kitchin, Gleeson and Dodge, 2013: 483).

Accordingly, the focus shifts from the issue of representation, and partially sidesteps more general questions of abstraction to ask exactly how maps effect, relate, enact and affect (Darling, 2014: 485). More specifically with respect to the question of temporality, it is a question of how maps are involved in 'generating spatio-temporal atmospheres' (Ash, 2013: 26). This implies not a general linear notion of time, but one specific to the encounter between the map and its territory. Time thus conceived, Bryant (2013: 20) argues, is counted among 'those properties that really do belong to things and the efficacy things organise on other things'. It originates in the encounter between the map and its territory; not simply as the result of their relationship as such, but through the map's particular 'capacity ... to move, threaten, inspire, and animate (Bennett, 2004: 358). Thus, the PathoMap, in its encounter with New York's metagenome, affects urban space such that it enacts a configuration of space and time. The question for an object-oriented cartography is to describe how exactly the map affects the urban and, consequently, what sort of time this involves. In particular, in this case of the PathoMap, it will explore to what extent it allows for a social imagination of the future.

## Urban informatics and the logic of preparedness

The aspiration for the PathoMap to approach ever more closely a real-time representation of the New York City microbiome is envisioned as a project of urban informatics. The latter, as the driver behind smart cities, is founded on the promise of the development and application of information technologies to render the city more transparent, legible and comprehensible. In New York, CUSP profiles itself as a world-leading research centre setting the agenda for urban informatics, pushing the boundaries of this emerging field. It promotes the developing of the capacity to generate and combine increasingly disparate sources of data in order to improve the efficiency of urban processes such as transport, energy use and waste management. In general, urban informatics employs 'large amounts of highly granular dynamic data' (Burke *et al.*, 2010: 328–331), which:

> include social networking sites, environmental systems and monitors, individual and institutional information sets of various kinds, RSS news feeds, and so on, and are understood as non-standard in urban terms.

In exploring a variety of techniques available to describe a certain bridge, for example, Burke *et al.* (2010) describe the challenge of successfully integrating heterogeneous sources of data. Insisting on a dynamic and adaptable 'ecology of techniques', this requires extensive processes of data collection, parsing, conversing, formatting and filtering. Increased computational capabilities are put to work to facilitate these processes in order to visualise the complexities of urban systems in real-time (Townsend, 2009: xxv). Ultimately, as these developments are intimately linked with city planning and management, this should push theories about the functioning of the city to focus on increasingly short-term time frames, probing for a 'thinking of cities as being plannable in some sense over minutes, hours and days, rather than years, decades or generations' (Batty, 2013: 276). As such the PathoMap project fits directly into the CUSP's (2014: no pagination) objective of 'developing cost effective risk management and emergency management practices that address preparedness, mitigation, response and recovery for both natural and man-made disasters to assure safety and security'.

This notion of preparedness is analysed by Fearnley (2007), who traces the history of US public health policies as part of the national civil defence system characterised by a logic of preparedness in contrast to that of insurance. While the latter, as Fearnley shows, informed the post-war establishment of social welfare states with national health care systems in Europe, the logic of preparedness in the US focuses on the early detection of emergence of diseases and the mitigation of effects. In shifting attention from the surveillance of people to the surveillance of diseases under the regime of pathogen preparedness, Fearnley (2005: 2) locates the development of a system of 'syndromic surveillance'. This system aims to identify correlations between different data sources, such as: the records of emergency room patients' symptoms, the sales of pharmaceuticals, and registers of emergency calls. By continuously comparing data, the syndromic surveillance system hopes 'to detect an unexpected, improbable epidemic in its emergent stages' (Fearnley, 2005: 29). To be sure, the PathoMap is meant to function in conjunction with a host of other, more traditional, forms of monitoring of disease. For instance, Afshinnekoo *et al.* (2015) are able to assure the public that the findings of large amounts of antibiotic-resistant bacteria, as well as some instances of anthrax and bubonic plague, are no real threats by putting these into a wider context. Yet what is important is that the PathoMap adds to such a regime of preparedness, extending the velocity of detection by not only relying on the correlation of data associated with disease, but to directly shift 'attention on the microbial agents themselves' (Fearnley, 2007: 5). This marks a shift from derivative, 'imperfect and often anecdotal' (Koonin, 2013: no pagination) information to an ambition of direct, perfect and total knowledge which typifies urban informatics as a science.

The rising importance of this concept of preparedness can be understood as part of a wider network of 'anticipatory action' arising from the increasing importance of new technologies in social and political life (Anderson, 2010). In the context of urban informatics, the formalisation and legitimation of anticipatory action as response to threats of terrorism, biosecurity and ecological disaster (Anderson, 2010) is routinely expressed in the field's self-proclaimed potential to deliver sustainable, efficient and resilient cities. These tendencies fundamentally reshape originally linear temporalities by bringing – in particular ways – the future into the present; rewriting, in the process, the role of the past. The specificity of this consists largely of the way in which uncertainty is dealt with, as 'the proliferation of anticipatory action, and the emphasis on an open future, is inseparable from a spatial-temporal imaginary of life as contingency' (Anderson, 2010: 780). This imaginary entails, Anderson (2010: 781) argues, an emphasis on a complex world consisting of an infinite multiplicity of flows and rhythms, and on life as 'unpredictable, dynamic and non-linear'. Such a spatial-temporal imaginary can also be seen to motivate projects such as the PathoMap; asserting the complexity of the world and proposing money, sewer systems and subway stations as primary locations of data collection, with a focus on the always lurking danger of possibly lethal pathogens.

## A fluid topology

Under urban informatics, the city – smart, sustainable, resilient – becomes a set of processes, monitored (preferably in real-time) through a wide range of (preferably automated) technologies (see also de Lange's chapter in this volume). These monitoring practices work together to establish a baseline to mark the data that counts as normal, against which the abnormal, exceptional and possibly disruptive can be recognised. This 'politics of possibility' (Amoore, 2013), does not count on a statistical extrapolation of a limited sample with clear temporal boundaries, but involves a continuous process of negotiation to demarcate normal from abnormal in order to enable a pre-emptive mode of intervention. If the monitoring of New York's microbiological life can be done increasingly automatically, this negotiation is what is at stake in the success of the PathoMap. As Chris Mason, project director, explains:

> of course, [Mason] notes, just because they identified certain bacterial species in the subways, 'that doesn't mean they're pathogenic,' and a big concern is igniting fear in the public. At this point, it's unclear how predictive the microbial data will be, and

'you don't want to have a panic attack,' he says. The first step, he adds, is simply to establish a baseline'. (Akst, 2013: no pagination)

Ideally, the highly granular detection methods of the PathoMap will be able to pick up emergence on the level of the individual microbe. However, the finding of 'strains of infectious bacteria such as Pseudomonas and Staphylococcus' in some of New York's subway stations (Akst, 2013: no pagination) becomes meaningful only if deemed to deviate from the norm. The PathoMap watches over the city's microbiome which is naturally in constant flux – viruses emerge and disappear. Metagenomic analysis of Penn Station, for instance, indicated strong variations of distinct types of microbes at different times of the day (Afshinnekoo et al., 2015). Yet only when these pass the threshold of the pathogenic – when it threatens the integrity of the city through a high-impact catastrophe – will it trigger the alarm to initiate necessary preventive measures to pre-empt the danger. In this sense, the timing of intervention becomes crucial: too late means that the city may be stricken by the disastrous effects of an infectious pandemic; too early, as Mason observes, and New York could be captured by an arguably even more infectious spell of panic.

As Afshinnekoo et al. (2015: 11) conclude in their evaluation, the achievement of a real-time monitoring system for the 'dynamics of the urban metagenome' will be to the benefit of large numbers of people. The vision of achieving an automated cycle of detection, evaluation and intervention enables the PathoMap to imagine the city as a fluid object. As a fluid object, constitutive elements and relations between elements may be added on or disposed of, but only 'gradually and incrementally' (Law and Mol, 2001: 614) if the coherence of the overall object is not to be disrupted. This is how continuous transformation establishes the invariance of the object in fluid space: 'so long as it flows there is the possibility that the transformations that it undergoes will not lead to abrupt changes' (Mol and Law, 1994: 658). In fluid space, the object is allowed to transform continuously, if only gradually and within a set of mobile and over-determinant boundaries. Here, change is inherent in the dynamics of the object – not subject to a linear past to be analysed through statistical methods focusing on discovering patterns to calculate probabilities. The relevance of the PathoMap hinges on the idea that New York City's microbiome is always in motion and it is precisely these dynamics which it aims to capture.

Mol and Law's notion of topology is deployed here to indicate how different spatial formations – i.e. heterologous topological spaces – carry with them concomitant conditions for the progression of time and the possibilities for change. In other words, fluid space is a conceptual framework that transcends a problematic opposition between space and time, 'to argue that they are integral to each other' (Massey, 2005: 47). The flow of time depends on the constant realignment of elements in space at the same time as the fluidity of space is conditional on the level of temporal openness. Massey's (2005: 48) charge that 'at a minimum, for time to be open, space must be in some sense open too' can thus be read backwards and forwards. This fluid imagination has a radical political potential, as it can offer a promise of a qualitatively different and hopefully better future. Whether or not the PathoMap fulfils such potential, however, depends on the nature of this fluidity. The crucial task for a *critical* object-oriented cartography is to assess to what extent it enables a topological imagination where the future is truly contingent and open to the unexpected. With the growing importance of all sorts of technological devices in society, it becomes essential to distinguish between those that gesture 'towards dynamics that are *internal* to socio-technical formations' and those that uphold more conventional and rigid conceptual separations between the social and the technological (Marres, 2012: 292, original emphasis).

## The pre-emption of possibility

The topologically distinct fluid space is the spatial form of the time of 'becoming' and 'emerging' as inherent to Anderson's (2010) 'life as contingency' paradigm, which insists on the inherent uncertainty and unpredictability of the modern, global world. The PathoMap is prepared for the omnipresent, impending danger of the terrorist living amidst the daily flow of commuters, or for the possibility that lethal microbes are already living in New York's sewers, subways, parks, or even people's wallets – for 'in the eyes of authority – and maybe rightly so – nothing looks more like a terrorist than the ordinary man' (Agamben, 2009: 23). In a world in which 'change cannot be understood as the linear outcome of past conditions or present trends', the prevention of pandemics works according to a logic where 'the causes of disaster are presumed to incubate within life' (Anderson, 2010: 782). Consequently, as the pre-emptive logic marks a shift from a politics of risk to a 'politics of the possible' (Amoore, 2013), 'disease surveillance as pathogen preparedness embodies a discontinuous or sporadic temporality of government' (Fearnley, 2007: 16). In this sporadic form of government, the PathoMap establishes a rhythm with the city from emergence to detection to intervention. In this way, it forms part of an urban assemblage in which both the map and the city figure as objects; such that the former is instrumental in the fluidity of the latter.

This fluidity is not the outcome of a pre-established assemblage, produced through the relationship between map and city, but is a result of the specific, non-relational capacity of the PathoMap to affect urban space. It is a direct result of the way the pre-emptive logic of anticipatory action that it enables, brings the future into the present. The focus on the omnipresent threat of pandemics marks the particular logic of anticipatory action enabled by the PathoMap project through its recourse to affect as '*component of passage* between mechanisms, orders of phenomena, and modes of power' (Massumi, 2005: 7, original emphasis). In this passage, decisions are no longer based on logical argumentation and calculative reason but on an affective charging of the empirical object. This is the map as device in its 'para-empirical condition' (Kurgan, 2013: 35), standing alongside the empirical body of the metagenome. It produces an increasingly seamless connection between the separate facets of 'bioterrorism threat detection, rapid-response, and containment' (PathoMap, n.d.: no pagination) in two steps. First, it offers direct and complete knowledge of the urban microbiome encompassing it temporally (i.e. in real-time) and spatially. This allows the map to appear at its most transparent, as if it were in fact the territory. Second, it is able to translate this knowledge straight into a sign of alarm so that the fear of a bioterrorist attack or the outbreak of an epidemic manifests itself, without the need for the threat to materialise:

> the identity of the *possible* object determines the affective quality of the *actual* situation. And that's a fact. Its quality has actualized, without the object itself materialising. It has taken affective passage from the future to the present, on the coattails of the time-inverse sign of alarm. (Massumi, 2005: 9, original emphasis)

Thus, the PathoMap becomes instrumental in turning – para-empirically – the detection of a particular form of microbiological life into a sign of alarm that transfers onto the actual, empirical object, the affective value of the virtual object of a bioterrorist attack. Parisi (2012), in her discussion of parametric design, puts this self-effecting mechanism of the affective decision in terms of a regime of topological control in order to think of anticipatory action as a specific way of structuring change and continuity. Through a system of 'generative programming of parameters', in which the real-time input of heterogeneous data sources are integrated and made to respond to each other, this form of design transforms the urban into 'one smooth machine of continual variation' (Parisi, 2012: 177). The 'discontinuous or sporadic temporality' observed by Fearnley corresponds to Massumi's (2005: 6–7) 'lightning brightness of the foregone conclusion'. The reason this conclusion becomes foregone is that it is based on the self-effecting mechanism of the affective fact, according to which the empirical fact as a sign of a possible future actualises the virtual affectively, without it needing to materialise in reality. Thus, a new politics of neoconservatism marks a 'tension between continued growth and becoming in the open neoliberal field of the capitalist system, and the sovereign closure of the foregone event' (Massumi, 2005: 7). Most importantly with regard to time, the decision becomes based not on the 'indefinite future of the what-may-come', but instead on the certainty of the '"will have" of the always-will-have-been-already' (Massumi, 2005: 6).

Here a picture emerges not of possible futures emergent in the present – an open future, life as contingency – but rather of a predetermined course of action inscribing itself into the past in order to bypass the present, denying it any significant claim to agency. As the turn to real-time monitoring of possible threats displaces the focus on statistical methods to calculate the probable based on past trends, the past is not merely disposed of as a factor in the decision-making process. Instead, as the empirical detection of microbial emergence makes its affective passage into the sign of danger, the eventual decision to intervene comes 'in the mode of having always been already' (Massumi, 2005: 6). As 'the future-past colonises the present' (Massumi, 2005: 6), the latter is sidestepped with the disqualification of logical reason's ability to explain the decision in a linear argument – valuable only for the extraction of empirical facts. Thus, the pre-emptive logic not so much brings the future into the present, but rather imposes the former in order to suppress the latter which, as a moment of discussion, calculation and deliberation, has to be avoided for its disposition to the weakness of uncertainty. In this way, the PathoMap becomes a typically modern apparatus in Agamben's (2009) terms by way of its fundamentally desubjectifying effect. That is, it is enlisted by a form of governmentality in which urban planning is rendered purely in terms of technical questions – 'a pure activity of government that aims at nothing other than its own replication' (Agamben, 2009: 22). This form of governmentality goes hand in hand with a science of cities in which urban planning as a set of 'integrated, quantitative, predictive, science-based' (Bettencourt and West, 2010: 912) techniques is opposed to traditional evidence-based policy-making, prone to all sorts of disagreements, values and opinions (see also Gleeson, 2013).

## Conclusion

To a degree, the project of the PathoMap certainly is imaginative: the creation and implementation of a pathogen weather map as a novel approach to syndromic surveillance testifies to a vivid imagination of the potential of new information technologies. Understood as a device (see for example Marres, 2012) it is able to mobilise a topological imagination, which establishes the city as a fluid object, not merely because of it being the object of a topological analysis, but because 'it is built into the technology itself' (De Laet and Mol, 2000: 225). The map then does not so much figure itself as a fluid object, a mutable mobile, but is instrumental in structuring reality in fluid terms. Accordingly, as a critical object-oriented cartography, it is important to evaluate the PathoMap by pointing towards the way in which the positing of the city as a fluid object by urban informatics enables a social imagination which opens up, or closes down, emergent possible futures. With respect to time, indicating a fluid topological imagination consolidated in the PathoMap fundamentally implies the assumption that 'the future will radically differ from the here and now' (Anderson, 2010: 780). Accordingly, in fluid space, object continuity is not so much maintained despite its transformation, but precisely because of it, 'for *fluidity generates the possibility of invariant transformation*' (Mol and Law, 1994: 658, original emphasis).

Yet, rather than appreciating the possibilities for continual, gradual change as a basis for invariant transformation, the PathoMap is devised to suppress uncertainty, keep novelty in check and guide it along the predetermined lines of the smart, resilient city, reassuring the public that the future will be fundamentally the same as the here and now. Rather than opening up the imagination to what the city could be, the PathoMap thus deploys a topological imagination of the fluid object in order to reinforce the grip on the possible according to established ideas about what the city should be. As such, it is part and parcel of a perspective which advocates the necessity of a 'grand unified theory of sustainability' in order to ultimately produce completely 'predictable cities' (Bettencourt and West, 2010: 912–913). Discussing the case of Songdo in South Korea, Halpern *et al.* (2013: 298–299) problematise precisely this politics of possibility as new epistemology of the emerging 'test-bed urbanism'. The real problem here is that 'when prediction collapses into production, we lose any possibilities of emergence, of change, or of dynamic life' (Halpern *et al.*, 2013: 299). What the earlier critical cartographers, such as Harley, understood was that the representation in a map can never fully account for the territory – it will always, by definition, be an abstraction, with the inevitable loss of some of its qualities as a consequence. In other words, central to constructivist critical cartography was the critique of representation for its abstractive tendencies always taking away from the complexity of the territory.

The shift towards an object-oriented critical cartography, in contrast, hopes to open up the question of how maps can add to rather than subtract from this vibrancy and, thus, how they can 'help us feel more of the liveliness' (Bennett, 2012: 232) of the places they represent. An object-oriented critical cartography explores the particularity of a map not to isolate it from its environment, but to more adequately understand its relationship to other objects and its place in the system. It considers the agency of the map itself so that it is neither a product of larger regimes of power, nor a neutral representation of its territory.

To be sure, object-oriented or 'thing-materialist' analyses are relevant to the study of maps of all times. Yet arguably, the insistence on the affective capacity of the map as object becomes even more pertinent with the abundance of digital maps and data visualisations. Several elements of this are important: the ability to incorporate increasingly large datasets and draw together ever more disparate elements; the progressively near real-time capacity to respond to empirical fluctuations; and the proliferation of devices and the diverse forms of interaction they afford. All of these add to the complexity of relationships between the map and the territory and, moreover, highlight the influence of maps and their potential to affect. An object-oriented approach is meant to pay attention to these expanding affective capacities and, critically, to assess the extent to which maps in particular contexts actually advance and invigorate understandings of this complexity. Finally, in the context of temporality, this may imply thinking about how, rather than commanding the pre-emptive elimination of uncertainty, maps can prompt the speculative provocation of possibility.

# References

Afshinnekoo, E., Meydan, C., Levy, S. and Mason, C. E. (2015) Geospatial resolution of human and bacterial diversity with city-scale metagenomics. *CELS*, 1(1): pp. 1–15.
Agamben, G. (2009) *What Is an Apparatus and Other Essays*. Translated by D. Kishik and S. Pedatella. Stanford, California: Stanford University Press.
Akst, J. (2013) Metropolome. *The Scientist*, December 1. [Online] Available at: www.the-scientist.com/?articles.view/articleNo/38376/title/Metropolome/ (accessed 3 May 2015).
Amoore, L. (2013) *The Politics of Possibility: Risk and Security Beyond Probability*. Durham, North Carolina: Duke University Press.
Anderson, B. (2010) Preemption, precaution, preparedness: Anticipatory action and future geographies. *Progress in Human Geography*, 34(6): pp. 777–798.
Ash, J. (2013) Rethinking affective atmospheres: Technology, perturbation and space-times of the non-human. *Geoforum*, 49: pp. 20–28.
Batty, M. (2013) Big data, smart cities and city planning. *Dialogues in Human Geography*, 3(3): pp. 274–279.
Bennett, J. (2004) The force of things: Steps toward an ecology of matter. *Political Theory*, 32: pp. 347–372.
Bennett, J. (2012) Systems and things: A response to Graham Harman and Timothy Morton. *New Literary History*, 43(2): pp. 223–233.

Bettencourt, L. and West, G. (2010) A unified theory of urban living. *Nature*, 467: pp. 912–913.
Bryant, L. (2013) Politics and speculative realism. *Speculations: A Journal of Speculative Realism*, 4: pp. 15–21.
Burke, A., Coorey, B., Hill, D. and Mcdermott, J. (2010) 'Urban micro-informatics: A test case for high-resolution urban modelling through aggregating public information sources'. In: Bharat, D., I. Kang Li, A. and Park, H. J. Hong Kong (eds) *New Frontiers: Proceedings of the 15th International Conference on Computer-Aided Architectural Design Research in Asia*, Hong Kong, pp. 327–336.
Campkin, B., Mogilevich, M. and Ross, R. (2014) How images shape our cities. *The Guardian*, December 1. [Online] Available at: www.theguardian.com/cities/2014/dec/01/picturing-place-how-images-shape-our-cities-snow-cholera-corbusier-graffiti (accessed 3 May 2015).
Crampton, J. W. (2001) Maps as social constructions: Power, communication and visualization. *Progress in Human Geography*, 25(2): pp. 235–252.
Crampton, J. W. and Krygier, J. (2006) An introduction to critical cartography. *ACME: An International E-Journal for Critical Geographies*, 4(1): pp. 11–33.
CUSP (Center for Urban Science and Progress) (2014) *Research: Disciplines, Domains, and Projects*. [Online] Available at: http://cusp.nyu.edu/research/ (accessed May 22 2014).
Dale, B. (2013) Mapping microbiotic life in New York City's guts. *Next City*, October 23. [Online] Available at: http://nextcity.org/daily/entry/mapping-microbiotic-life-in-new-york-citys-guts (accessed 3 May 2015).
Darling, J. (2014) Another letter from the Home Office: Reading the material politics of asylum. *Environment and Planning D: Society and Space*, 32: pp. 484–500.
De Laet, M, and Mol, A. (2000) The Zimbabwe bush pump: Mechanics of a fluid technology. *Social Studies of Science*, 30(2): pp. 225–263.
Fearnley, L. (2005) From chaos to controlled disorder: Syndromic surveillance, bioweapons, and the pathological future. *ARC Working Paper 1*. [Online] Available at: http://anthropos-lab.net/wp/publications/2007/01/fearn_chaos_to_disorder.pdf (accessed 22 May 2014).
Fearnley, L. (2007) Pathogens and the strategy of preparedness. *ARC Working Paper 3*. [Online] Available at: www.anthropos-lab.net/workingpapers/no3.pdf (accessed 22 May 2014).
Gleeson, B. (2013) What role for social science in the 'urban age'? *International Journal of Urban and Regional Research*, 37(5): pp. 1839–1851.
Halpern, O., LeCavalier, J., Calvillo, N. and Pietsch, W. (2013) Test-bed urbanism. *Public Culture*, 25(2): pp. 272–306.
Harley, J. B. (1989) Deconstructing the map. *Cartographica*, 26(2): pp. 1–20.
Harman, G. (2012) The well-wrought broken hammer: Object-oriented literary criticism. *New Literary History*, 43(2): pp. 183–203.
Hayles, N. K. (2014) Speculative aesthetics and object-oriented inquiry (OOI). *Speculations: A Journal of Speculative Realism*, 5: pp. 158–179.

Hookway, B. (2014) *Interface.* Cambridge, Massachusetts: The Massachusetts Institute of Technology Press.
Kitchin, R. and Dodge, M. (2007) Rethinking maps. *Progress in Human Geography*, 31(3): pp. 331–344.
Kitchin, R., Gleeson, J. and Dodge, M. (2013) Unfolding mapping practices: A new epistemology for cartography. *Transactions of the Institute of British Geographers*, 38(3): pp. 480–496.
Koonin, S. E. (2013) The promise of urban informatics. *Center for Urban Science Progress*, May 30. [Online] Available at: http://docplayer.net/5866972-Center-for-urban-science-progress-the-promise-of-urban-informatics.html (accessed 1 August 2015).
Kurgan, L. (2013) *Close Up at a Distance: Mapping, Technology and Politics.* New York: Zone Books.
Law, J. and Mol, A. (2001) Situating technoscience: An inquiry into spatialities. *Environment and Planning D: Society and Space*, 19(5): pp. 609–621.
Law, J. and Ruppert, E. (2013) The social life of methods: Devices. *Journal of Cultural Economy*, 6(3): pp. 229–240.
Marres, N. (2012) On some uses and abuses of topology in the social analysis of technology (or the problem with smart meters). *Theory, Culture and Society*, 29(4/5): pp. 288–310.
Massey, D. (2005) *For Space.* London: Sage Publications.
Massumi, B. (2005) The future birth of an affective fact. *Conference Proceedings*: Genealogies *of Biopolitics.* [Online] Available at: http://browse.reticular.info/text/collected/massumi.pdf (accessed 1 May 2015).
Mol, A. and Law, J. (1994) Regions, networks and fluids: Anaemia and social topology. *Social Studies of Science*, 24(4): pp. 641–671.
New York University (2013) NYU announces winners of 'Grand Challenge' science competition. *NYU News*, September 23. [Online] Available at: www.nyu.edu/about/news-publications/news/2013/09/23/nyu-announces-winners-of-grand-challenge-science-competition.html (accessed 22 May 2014).
Parisi, L. (2012) Digital design and topological control. *Theory, Culture and Society*, 29(4/5): pp. 165–192.
PathoMap (n.d.) *PathomapAbout.* [Online] Available at: www.pathomap.org/about/ (accessed 22 May 2014).
Perkins, C. (2004) Cartography – cultures of mapping: Power in practice. *Progress in Human Geography*, 28(3): pp. 381–391.
Townsend, A. (2009) 'Foreword'. In: Foth, M. (eds) *Handbook of Research on Urban Informatics: The Practice and Promise of the Real-Time City.* Hershey, Pennsylvania: Information Science Reference, pp. xxiii–xxvii.
Wood, D. (1993) What makes a map a map? *Cartographica*, 30(2/3): pp. 81–86.

# 11

## From real-time city to asynchronicity: exploring the real-time smart city dashboard

*Michiel de Lange*

### A plea for asynchronicity

In a thought-provoking 'design fiction' exercise, design researchers Bleecker and Nova invert the discourse of instantaneity in urban computing and digital cartography (Bleecker and Nova, 2009). Urban new media tend to promote a speeding up of time:

> there is here a conspicuous arms race towards more instantaneity, more temporal proximity between events, people and places. Communication is promoted to be 'just-in-time'; feedback to your activities should be in 'real-time' as if you were playing a video-game character. Speed is essential, and this never-ending battle with time – to eliminate it – makes things happen instantaneously. (Bleecker and Nova, 2009: 29)

By elevating 'real-time' to a prime design objective, urban new media in fact are geared to dispose of time – in the sense of duration – as a limiting factor altogether. Bleecker and Nova forward the notion of *asynchronicity* to explore how urban computing technologies might afford more diversified interactions than the efficiency-driven real-time model:

> there are many geographies, asynchronous because we have individual experiences of the world. Fixed things become flows, and flows become the fixed point of reference … Perhaps we learn from this that computing in an urban setting should first of all not be about data and algorithms, but people and their activities. What happens when time everywhere is not synchronised, when it floats and lags a bit? (Bleecker and Nova, 2009: 19)

Bleecker and Nova argue that out-of-sync mapping reinserts serendipity into urban life by stimulating unexpected encounters. Asynchronicity allows citizens to appropriate the city by creating incremental maps that allow them to narrate their personal and collective 'sense of place'. The concept of asynchronicity also draws attention to the fact that systems tend to break and that underneath the myth of smooth 'always-on' availability of information there is the everyday messiness of technologies failing. Asynchronicity could be about doing something in a place and then getting back there to see what happened afterwards. This stimulates engagement with that place and other people. Bleecker and Nova (2009) refer to the Japanese urban game *Mogi*, where players had to hunt for treasures in the urban landscape but instead unexpectedly ended up using it as a social networking tool.

It remains, however, somewhat unclear what exactly the notions of real-time and asynchronicity mean and what their implications are. Bleecker and Nova make an implicit argument for a kind of 'slow mapping', yet the question remains: who or what needs to slow down, and what could be the implications of this? Is it about managing our lives at work, home, travel and so on in the slow lane? Do urban services need to be delivered more slowly? Is it a design imperative to create 'slow' situations and experiences, and what would that entail?[1] Bleecker and Nova note that the real potential of locative media and location-based services is 'the ability to find oneself relative to everywhere else' (Bleecker and Nova, 2009: 17). It is this people-centric relational view that I want to pursue here.

## Urban dashboards as interfaces to the smart city

The recent proliferation of work about urban dashboards seems to be almost canonising the field (Ciuccarelli, Lupi and Simeone, 2014; Kitchin, 2014; Batty, 2015; Holden and Moreno Pires, 2015; Kitchin, Lauriault and McArdle, 2015a; 2015b; Mattern, 2015; Wilson, 2015). Frequently invoked ancestors include: the automobile dashboard, the airplane cockpit, the space mission control centre, the financial boardroom and state-led industry monitoring. An iconic early urban dashboard was the Cybersyn control centre developed by the cyberneticist Stafford Beer for the Allende government in Chile in 1970 (Medina, 2006; Morozov, 2014; Batty, 2015; Mattern, 2015).[2] The left-wing government nationalised many private companies and had the Cybersyn management cybernetics system developed; a system that 'would network every firm in the expanding nationalised sector of the economy to a central computer in Santiago, enabling the government to grasp the status of production quickly and respond to economic crises in real-time' (Medina, 2006: 572). Due to technical difficulties, the system only allowed companies to transmit data once a day and reminding workers to manually do so proved a source of frustration for the project team (Medina, 2006: 587). Chairman of the Centre for Advanced Spatial Analysis Michael Batty notes regarding the Cybersyn project that 'the data were always out of sync' because the age of real-time computing had not yet ushered in (Batty, 2015: 29). By contrast, he insists that today 'the most rudimentary of dashboards applicable to displaying the routine operation of the city do collect data in real-time that are comparatively neutral in their factual complexion' (Batty, 2015: 30). This epistemological claim is problematic and needs unpacking.

First, it is helpful to make a provisional typology of the city dashboard. Categories and subdivisions below – neither comprehensive nor static – are based on: the type of platform used to map and share data, the kind of data or indicators and what can be done with them, the way in which information is visualised, and the purpose for which they are being used.

## Platform

Real-time data can be delivered on various platforms. First, centralised physical control rooms are used for managing city processes and operations. Cybersyn fits here, as well as IBM's control room in Rio de Janeiro (Mattern, 2015).[3] Second, there are online platforms that display one or multiple processes.[4] Third, distributed mobile platforms deliver real-time data, for instance to enterprise employees, police squads, quantified selfers engaging in sustained self-tracking, and consumers.[5] Such platforms may be either open to the public – e.g. installed in public places or available online – or closed.

## Data and uses

Another subdivision consists of the kind of urban data or indicators and what can be done with them. Kitchin, Lauriault and McArdle (2015a) distinguish between single indicators or composite indicators that combine measurements. Second, they identify various uses: *'descriptive or contextual indicators'* used for insights; *'diagnostic, performance and target indicators'* used to assess performance; and *'predictive and conditional indicators'* used to anticipate future situations (Kitchin, Lauriault and McArdle, 2015a: 8–9, original emphasis). We may further differentiate between standardised and comparable data and local situation-specific data; quantitative data (e.g. temperature) and qualitative social and cultural data (e.g. levels of happiness) (Kitchin, 2014); user-generated data in often distributed ways and data generated by professionals and often kept in a central repository.⁶ At the legal level, data may be open source or closed/proprietary data. For example, some projects allow raw data to be downloaded and/or use open APIs for others to reuse data (e.g. oscity.nl), while others are closed (e.g. real-time policing platforms).

## Visualisation

We may also differentiate between ways of displaying information on dashboards, via charts and graphs, diagrams or maps (Kitchin, Lauriault and McArdle, 2015a). Distinctions can further be made between indexical interfaces, with a direct reference to the measured object (e.g. a gas tank meter), and symbolic interfaces that involve a further translation (e.g. an alarm light indicating an empty tank) (Mattern, 2015), between fixed and interactive visualisations (Kitchin, Lauriault and McArdle, 2015a), and between dynamically refreshing or cumulative ones.

## Purpose

The intended purpose of urban dashboards and the associated agency attributed to stakeholders may vary widely, although in practice there is overlap. We can distinguish between dashboards providing accountability and legitimacy through transparency (Perez and Rushing, 2007: 11), collecting intelligence and providing cues for action, managing emergencies, benchmarking and comparison, surveilling and controlling (Kitchin, 2013: 15), offering democratising tools for civic empowerment and social change (Holden and Moreno Pires, 2015), and providing creative opportunities to hackers, artists, app makers and citizens, like creating data-narratives (de Waal and de Lange, 2014).

Like any media technology, real-time urban dashboards 'do not reflect the world as it actually is, but actively frame and produce the world' (Kitchin, Lauriault and McArdle, 2015a: 24). The asserted power of dashboards is that city managers (and citizens to a variable degree) receive instant, transparent and realistic information about urban processes under the presumption that one can 'know' the city *as it actually is*' (Kitchin, Lauriault and McArdle, 2015a: 16, emphasis added). Urban dashboards have been critically questioned on the basis of their ontological, epistemological and political assumptions. On the one hand, dashboards may open up data to public consumption and use, yet on the other hand they cultivate a top-down, technocratic vision (Ciuccarelli, Lupi and Simeone, 2014: Mattern, 2015). Real-time mapping and dashboards provide a powerful realist epistemology (Kitchin, Lauriault and McArdle, 2015a; Mattern, 2015). An issue is the validity of the data. Mattern notes that the target audience 'likely has only a limited understanding of how the data are derived' yet will base actions on them unquestioningly (Mattern, 2015: no pagination). Similarly debatable is the assumption that cities 'consist of a set of knowable and manageable systems that act in "rational, mechanical, linear and hierarchical" ways' (Kitchin, Lauriault and McArdle, 2015a: 14). This realist epistemology drives a 'new managerialism' (Kitchin, 2013; Kitchin, Lauriault and McArdle, 2015a; see also Morozov, 2013). Decision-making is based on what can be measured and quantified – purported hard and realistic data. But what data are left out? The messy data that cannot be neatly quantified and visualised, Shannon Mattern argues (Mattern, 2015). Moreover, the politics of 'qualculation' (Callon and Law, 2003; Thrift, 2008: 24) runs counter to making decisions based on careful and necessarily slow rational deliberation or even on affect and emotion, the 'gut feeling' that have recently been resuscitated as an important driver of our actions (Gladwell, 2005; Levitt and Dubner, 2005; Ariely, 2008).[7] Affect and emotion, which have become more prominent in both urban studies and in computer research, remain largely absent from smart city visions of what makes a city liveable (for a discussion see de Lange, 2013).

## Time and the dashboard

From this discussion of dashboards, we can proceed to unpack the oft-invoked real-time adjective and attempt to modify the discourse by highlighting asynchronicities. To do so, I build on sociologist Barbara Adam's identification of seven elements constitutive of time (Adam, 2008). 'Temporal frames are not given but chosen', she asserts (2008: 2; see also Adam, 1990). This helps to unbox what is presented as real-time mapping beyond the merely discursive level, and aids in developing asynchronicity further as an alternative heuristics to scrutinise urban dashboards as a way to govern today's cities.

### 1 *Time frame*

The first of Adam's structural features of time is the '*time frame*', which refers to a bounded unit with a beginning and an end, like a day, a year, a life time, a generation, or an epoch (Adam 2008: 2, original emphasis). Real-time dashboards are never really real-time. Information sampling and mapping ideally occupies an infinitely small time frame yet always has a certain 'refresh rate'. Technically, the encoding/decoding of digital information in bits and bytes occurs in discrete units, for example the sampling rate (time slice) and the bit rate (resolution) in encoding digital music. Time frames are also involved in the algorithmic processing of information. Any digital processing involves latencies incurred by among others memory buffering, CPU scheduling and process interrupts. Zero-latency is always an approximation. Furthermore, the actual visualisation can be temporally framed, for example the number of frames per second or when information is display cumulatively and the counter is reset at some point. Finally, people's response time to information and their potential ensuing actions take place within a certain time frame. Notions like real-time, immediacy, liveness and transparency blackbox the temporal dynamics between sensing and displaying. Asynchronicity – understood here as a measure of *latency* – raises questions about the rhetoric and subjective experiences of real-time. Within what time frame do people perceive information as immediate? What is an acceptable time frame for automated responsiveness? In line with Bolter and Grusin's argument about the 'double logic of remediation', prevailing claims to real-time involve both the erasure and the multiplication of media (Bolter and Grusin, 2000: 5). It is precisely the whiz-bang interfaces, multiple screens, rapidly alternating maps, graphs and stats in flashy colours and smooth effects of smart city dashboards that provide the suggestion of direct and transparent access and power over urban processes. Asynchronicity thus shifts our attention to the question how the design and medium-specific qualities of urban dashboards constitute the perceived governability of smart cities.

## *2 Temporality*

*Temporality* asks how time unfolds and what direction it takes. It involves a procedural view of time as changing, ageing, growing and irreversible (Adam, 2008: 2). Real-time dashboards attempt to capture events transpiring as they happen, in an immediate 'now'. It has little eye for temporality. Real-time risks reducing events to isolates, to singularities. Algorithms identify the out-of-the-ordinary and at the same time routinise the responses to them. The exceptional becomes homogenised. Change is much harder to grasp. Indeed, Michael Batty notes that so far there has been only limited progress in terms of moving 'away from real-time monitoring of performance to some more abstracted interpretation of how the state of a city is changing over the longer term' (Batty, 2015: 31). The obsession with real-time, in my view, precludes more complex temporalities – e.g. from identifying correlations and patterns to weaving actual stories out of isolated events – that complement our understanding of urban dynamics. Michael Flowers, the former chief analytics officer of New York City, describes how data from different city agencies were integrated to make correlative fire-hazard analyses and risk-filters based on intelligence about illegal housing conversion (Flowers, 2013). This allowed officials to prioritise the inspection of potentially dangerous locations. The question remains whether these predictive analytics address the far less visible and more tenacious issues of immigration and housing shortage. Elsewhere, Flowers is quoted as saying: 'I am not interested in causation except as it speaks to action. Causation is for other people, and frankly it is very dicey when you start talking about causation … You know, we have real problems to solve' (Morozov, 2014: no pagination). It would not be impossible to do such a thing if real-time city data analytics take the relationships between events as the central unit.[8] Temporal relationality is crucial and asynchronicity – understood here as temporal unfolding and recurrence – helps to consider how out-of-sync events refer to one another but do not coincide. Temporal relationality stretches time and folds it back onto itself, creating connections and narratives out of disparate events. To illustrate this somewhat abstract point, a recent study suggests that designers of persuasive technologies (e.g. quantified self-tracking apps) falsely rely on the 'egocentric loop' (Balestrini, 2013: no pagination). The process from data collection, to visualisation, to self-awareness, does not smoothly nudge people towards actual behavioural change. Instead, intermittent feedback via affective relationships with other agents might be more helpful. Through vicarious social interactions and recursions, more meaningful relationships may arise with one's own data, and in due time, actual change may unfold.

## *3 Timing*

The moment of something happening is a matter of *timing*, that is, taking place at a specific time. Adam notes that many aspects of our lives are synchronised: clock time, body time, seasons and climates, social time (e.g. opening hours), task-related timing, and timing associated with using specific communication technologies (Adam, 2008: 3). Social synchronisation and questions about right or wrong timing for coordinated actions are context-dependent (Adam, 2008: 3). Timing thus is a highly normative notion. The 'just in time' of real-time mapping is unquestioningly equated with being 'right on time'. What's more, dashboards increasingly employ predictive algorithms to map potential futures in anticipatory ways. Examples of predictive mapping and pre-emptive action include: crime mapping, spotting potential terrorist attacks and detecting fraudulent patterns (Crang and Graham, 2007; Crandall, 2010). Timing subtly shifts from what happens now to what might happen, a kind of future forward asynchronicity. This raises various questions, such as: when is the time right to act towards an anticipated future? How can a potential future become an actual present context? How desirable is it when human agency is bypassed? Where does this leave the opportunity for action by the smart citizen? The idea that now is always the 'right time' to act is an extremely technocratic notion. Real-time cybernetics, in many ways, runs counter to democratic politics, as a deliberately slow-acting counterweight to whimsical and over-hasty decisions, as well as counter to the practical view of good governance at quite a few city halls as the art of doing as little as possible.

## 4 Tempo

*Tempo* is about the speed at which something happens, the pace, rate of change and intensity. Tempo is often politically charged: who decides tempo for whom and why, and what frictions and clashes might occur? What expectations do people in different social domains have of certain activities in a certain time frame (Adam, 2008)? What happens, Adam wonders, when the speed of one domain like the internet percolates into other realms like family life? Speed and acceleration have been dominant themes in theorising metropolitan life, particularly in relation to emerging transport and communication technologies (de Lange, 2010). Scholars and commentators have looked at the speeding-up of the 'urban metabolism' (Townsend, 2000), and the new social and mental attitudes induced by rapid, plentiful and sustained information. Urban designers attempt to create legible cities that mitigate the risk of spatial disorientation and the looming psychological fear of getting lost (Lynch, 1960). Acceleration is a closely related trope. The current obsession with data governance parallels Williams and Srnicek's description of the neoliberal capitalist demand for acceleration and its claim to usher in an age of technological singularity (Williams and Srnicek, 2013). Paradoxically, urban dashboards are touted as providing stability and control while their added value for urban governance relies on ever-increasing acceleration in delivering information (Williams and Srnicek, 2013). Acceleration becomes an end in itself – with near real-time input for reactive or pre-emptive action – instead of a means towards better understanding our cities and empowering people's creativity. Creativity, asynchronous and slow in comparison to computerised systems, becomes ballast rather than a resource. Real-time technologies aspiring to infinitely speed up their own working quite literally preclude the *latent* potential of people to hack these technologies and use them for truly democratic collective self-mastery, governance and creation (Williams and Srnicek, 2013).[9]

## 5 Duration

*Duration* asks how long. It is the temporal equivalent of distance and its opposite is instantaneity. Duration, like distance, is a subjective and affectively charged notion. Instantaneous real-time dashboards map information in an eternal 'now' without a sense of duration. Asynchronicity – taken here as referring to an extended stretch of time between an event, its representation and a possible action – would allow for the development of an affective and emotional relationship to an indicator that is being mapped. For example, a recent study in the field of decision-making psychology shows that people are more willing to drink recycled waste water when it has been stored in an aquifer for ten years, or when it had travelled a hundred miles instead of one mile. New technologies can clean water in five minutes, however, this near-instantaneity runs out of sync with the perceived 'spiritual contagion' of recycled waste water as something that needs time to become purified and 'natural' again (Rozin *et al.*, 2015: 56–57). Asynchronicity seems key in such truly transformative processes, shifting from pollution to purity, from dangerous to safe, and so on. In this case, it also emphasises the magical-thinking involved in many profound decision-making processes; that which seems to escape neat logic and rationality and needs due time to evolve.

## 6 Sequence

*Sequence* is about the order of things, about succession and priority. The absence of a sequence of events means they are conflated into simultaneity. A typical real-time urban dashboarding sequence involves continual capturing of source data as input (the measured thing or event), the algorithmic processing of data along multiple steps, for instance prioritisation, classification, association and filtering (Diakopoulos, 2014), and feeding the output back to an interface and/or decision-making agent.

Klauser, Paasche and Söderström describe this sequence as: (1) generating, gathering and processing data derived from digitised urban systems; (2) interconnecting and fusing various data about everyday life; (3) data analytics (Klauser, Paasche and Söderström, 2014: 869–870). In the attempt to do all this in real-time, the steps in the process and their order and priority are frequently obfuscated. If we think of urban dashboards through the lens of asynchronicity – taken in this case as separate steps that do not neatly fall together – we are drawn to raise important questions, such as what the order of processing entails, what the consequences of this sequence are, to what extent this is open to scrutiny and modification, and who are allowed to design or exert influence on this at what moment. Batty, for instance, notes that the 'manual override' is a major feature of urban interfaces (Batty, 2015: 30). As automated systems become normative forces, it would be very interesting to find out the actual dynamics of overriding the system. The lens of asynchronicity – taken here as indicating a consecutive order of separate phases – helps us to question the veracity of what is being measured and displayed by contextualising every step along the way as a situation. For instance, in gathering data, we must pay attention to the fact that experiences, desires and behaviours can vary widely in time, e.g. between day and night and between the seasons. Generic cookie-cutter algorithms trigger instant action without taking temporal situatedness into account.[10]

## 7 Temporal modalities

*Temporal modalities* refer to the question when something happens, in the past, present or future (Adam, 2008: 2). Adam distinguishes between two standpoints towards the future: the future present and the present future (Adam, 2008). The present future approaches the future from the present, as 'mine to shape and create' (Adam, 2008: 7). The future present approaches present actions as seen from the future: the impact of present actions for future generations. This is the area of ethics. The first seems more individualistic, the other more collective. The real-time dreams of the smart city have a tendency to scoop up the future right now. It is potentially disenfranchising as it ignores the realm of ethics and the active role of citizens in shaping their city for posterity in sometimes conflicting or contradictory ways. Furthermore, Klauser, Paasche and Söderström (2014) observe how automated governing through code alters the relationship between past, present and future. Real-time regulation draws the different temporal modalities into a co-present state: it relies on designed and written code that is modelled on an analysis of the past and applied to the present to anticipate the future (Klauser, Paasche and Söderström, 2014: 877). Furthermore, they argue that governing through code performs the future, since code does not merely describe and analyse the present but also provides a grammar for action (Klauser, Paasche and Söderström, 2014). Asynchronicity here means inserting deliberate breaks into temporal modalities to learn from individual or collective displacements, for example by looking back in hindsight to continually reconstruct the past and learn from it, and looking forward into the future to understand how present actions matter.

From the above we see that asynchronicity always places something relative to something else, while real-time inflates relationships to an eternal here and now. The notion of asynchronicity creates room for understanding the relations and overlaps, so central to life in the city. It creates frictions and seams, rather than polishing them away into a smooth, seamless, efficient and optimised space. In the words of Adam, careful attention to time is needed to understand how 'the individual, the social, the institutional, the historical and the socio-economic, political and socio-environmental aspects of our lives are interconnected as well as mutually implicating and forming' (Adam, 2008: 4).

## The asynchronous smart city

Our breakdown of the notion of real-time according to Adam's typology and the lens of asynchronicity has teased out a variety of issues concerning real-time dashboards. In this concluding section, I highlight some elements that stand out: questions about representation, knowledge and politics.

The 'real' in real-time dashboards suggests a direct and unmediated view of the city as it really is, turning 'epistemology into ontology' (Mattern, 2015: no pagination). Furthermore, indicators are translated into actions. Dashboards, however, detach and decontextualise decision-making from the actual situation to which it pertains. 'Management at a distance' (Klauser, Paasche and Söderström, 2014: 870) – the spatial and social distancing between representation and action – derives its validity from the claim to temporal synchronisation between its various actors and operants. Impacts of these actions are only fed back insofar as they can be quantified and calculated as well. For quite some time, urban scholars and planning professionals have deployed media technologies to better understand cities. William H. Whyte for his long-running *Street Life Project* made the film documentary *The Social Life of Small Urban Spaces* (1979), in which he captured and analysed spatial and behavioural patterns of urbanites, combining a perspective from above with detailed on the ground observations. A more recent example is the use of GPS to track people's mobility patterns at the Technical University Delft (Van der Spek *et al.*, 2009). In these cases media technologies merely provide an added perspective of city life. But what happens when a technologically mediated view based on quantifiable indicators takes over and becomes the prime gateway to knowing about today's cities? Can everything be measured and quantified, what is being left out?

At the level of representation, asynchronicity draws attention to citizen-driven reprogrammability as a distinct feature of today's 'smart' cities. The modernist city was spatially preprogrammed to be used for single purposes: living, working, leisure, mobility, meeting (Hannerz, 1980). A city park equipped with WiFi (preferably with decent benches and good coffee) becomes an outdoor working place. Collaborative consumption platforms allow private goods to become pooled and collectivised. Urban facades allow outdoor spaces to become canvasses for artistic uses. Location-based urban games turn the city streets from a mobility infrastructure into a gameboard for leisure activities. In tackling pressing urban issues, being smart may in fact mean repurposing instead of reinventing the wheel. Innovation springs from these glitches in neat legibility. Mapping practices have evolved from making the city legible (Lynch, 1960), to writing the city with one's own experiences on annotated maps (Greenfield and Shepard, 2007), to contribute to what Marc Tuters and I – taking the UNIX metaphor a step further – have called 'executable urbanism' (Tuters and de Lange, 2013: 49). Action maps and participatory mapping practices can make the city 'hackable' for its inhabitants, that is, open to systemic change by everyone. Some of the ways in which this occurs are by providing an understanding of its inner workings, engaging people with the ongoing process, providing people with a horizon for action, and building collectives around shared issues of concern (de Lange and de Waal, 2013).

The underlying 'neat' systems perspective of the urban dashboard leaves out the messiness of mapping as a form of representation that necessarily involves generalisations, distortions, ambiguities, 'white lies' and so on (Walker, 2011: 63). Real-time mapping merely parses input into output. It precludes a learning trajectory. From the perspective of user centred design, instant gratification erodes the concept of value or what it takes to achieve something. Renaissance painters experimented with *anamorphosis*. They painted distorted images or elements that could only be seen from certain perspectives or with optical aids like cylindrical mirrors. *Anamorphosis* immobilised the onlooker to a particular vantage point (Bouman, 2013). Interesting for our discussion is that the technique deliberately aims to postpone understanding. Drawing attention to the medium itself, *anamorphosis* engages the spectator in a puzzle game to slowly expose secret layers of representation and meanings. Asynchronicity as a design parameter allows for a deferred sense of accomplishment in learning, feeling smart by piecing together an image out of the inherent distortions of the real-time map. The 'transparency' implied in dashboards means the opposite of experiencing the city viscerally, including its layers of secrecy, which can only slowly be unpacked yet never fully understood.

Second, I briefly dwell on some political implications. At the level of everyday politics, the real-time city appears a machine for delivering frictionless services. This runs counter to ideals of the city as a place of friction and dealing with otherness. At the level of politics as organised decision-making, asynchronicity highlights how smart city interfaces may provide horizons for action in spatial decision-making. Asynchronous maps, such as open data aggregator OSCity.nl, are always unfinished. They show their shortcomings and allow for ongoing processes of adaptation and mutation. They provide scripts rather than scenarios: affording diverging, iterative and open-ended play instead of singular, predefined, top-down planning narratives.

If we think of politics as participation, we observe a tension between competition and collaboration in smart city governance. *Real-time* is used as the equivalent of *smart*, individual or collective.[11] Many smart apps frequently resort to competition. Klauser, Paasche and Söderström (2014), for instance, describe how an IBM *Smarter Energy* Executive believes that people can be encouraged to change their behaviour based on real-time feedback about other people's energy patterns, which fosters a competitive desire to optimise electricity consumption (Klauser, Paasche and Söderström, 2014: 878–879). At best they leverage collaborative citizen creativity to aid in governing the city rather than assuming a veritable new role as city hackers that challenge this system.[12] Hackability as an affordance of truly smart cities means breaking out of the neat confines of urban governmentality and opening up new possibilities for (re)use of infrastructures and services, and allowing for unsolicited, clever, collective citizen initiatives (see Wakefield and Braun, 2014).

Lastly, politics is a question of fair redistribution and including the excluded. There are places that are not in-sync across various domains, like: law, politics, media, social institutions, or with the rest of the world. While real-time suggests everything is included, in every place there are political vacuums, asynchronous blank spots on the map.[13] We need to be aware that all cities, or parts of some cities, are not algorithmically governed equally.

## Conclusion

Maps have enabled us to imagine and relate to places outside of our immediate surroundings. The 'liveness' of today's media connect us to global geographies. What seems to be lacking is a sense of long-term global temporalities.[14] In this contribution, I have attempted to approach real-time urban dashboards through the alternative notion of 'asynchronicity' as a way to reflect on temporality in mapping the city. Asynchronicity highlights latency, recurrence, deferred understanding and imperfections of the mediating process. It opens up a long-term, slow perspective of the future and a more citizen-centric view of the smart city.

The urban dashboard stands in a long tradition in which ideals of transparency and directness pervade discourses about communication (see for example Peters, 1999). In a sense, the real-time urban dashboard reconciles the space and time-transcending capacities of information and communication technologies with place-based, proximate and co-present urban life. The dashboard serves to solidify a still uneasy affair between the two and their differences in the spatio-temporal modalities of the localised actionable here and the dispersed informational elsewhere (see Rodgers, Barnett and Cochrane, 2014).[15]

Future research may ask how the real-time city shapes subjectivity and identity though reflexive monitoring and the politics of quantification. The real-time city figures citizens as consumers seeking instant satisfaction or, as Jennifer Gabrys notes, as productive sensing nodes (Gabrys, 2014). How is our sense of self affected by our awareness of living a quantified life that is governed in anticipatory ways? To what extent can the eternal 'now' be reconciled with the careful self-reading and narrative emplotment that figures so prominently in many influential theories of identity (for example, Giddens, 1991; Ricoeur, 1992)? To what degree are transparency and univocality antithetical to heterogeneity, pretence and conceit, ambivalence and multiplicity that underlie personal and cultural identities in the city? How are subjectivities constituted through real-time modes of surveillance? Another line of research would be to focus on the actual design and production side of real-time mapping, for example by conducting ethnographic observations of actual processes of mapping and decision-making in smart city control rooms.

## Acknowledgements

An early version of this work was presented during the workshop 'Time travelers: Temporality and mapping', University of Oxford, 27–28 May 2014. The author wishes to thank the participants for their comments and suggestions.

## Notes

1 Their plea for slowness ties in with a recent call by Christoph Lindner for the slow smart city (Lindner, 2013). In Italy, *città slow* are positively conceptualised as prioritising quality of life over quantity of goods and services.
2 See the research and documentation website at: www.cybersyn.cl/ingles/cybersyn/ (accessed 4 December 2017).
3 See: http://cor.rio/ (accessed 20 December 2017).
4 An example is the London City Dashboard: citydashboard.org/london/ (accessed 4 December 2017).
5 An early example of a real-time city B2C dashboard was CitySense, which allowed its user to see what venues other users with similar consumer profiles were frequenting. An example of a mobile enterprise dashboard is Apple and IBM's *MobileFirst*: www.ibm.com/mobile (accessed 20 December 2017).
6 For more fine-grained typologies of data, see Kitchin (2014).
7 The urban dashboard is frequently described in terms of being the city's brain or operating system, in a striking parallel to popular research that sees the human brain as a supercomputer driving behaviour, experience and thought. This stands in stark contrast

to research that focuses on affect as a relational and distributed kind of 'smartness' (Zeidner, Matthews and Roberts, 2009; Arvidsson and Colleoni, 2012; Buser, 2014).

8   Kitchin points out that *relationality* is actually a defining feature of big data (Kitchin, 2014: 68). But this seems to be equated merely with dataset *compatibility* and database *connectivity*.

9   Antonio Negri in a response to the Manifesto proposes 'latency' as a potentially revolutionary force: www.e-flux.com/journal/53/59877/reflections-on-the-manifesto-for-an-accelerationist-politics/ (accessed 4 December 2017).

10  For example, an open fire hydrant may be cause for concern during wintertime but may develop into a great community event during a heatwave when left untouched for a while.

11  For example, the Waze real-time traffic app payoff is: 'Outsmarting traffic, together': see www.waze.com (Hind and Gekker, 2014).

12  Currently I am involved in a series of research projects called 'The Hackable City'. See http://themobilecity.nl/projects/amsterdam-hackable-metropolis/ (accessed 4 December 2017).

13  This is an insight I took from DEAF 2014, where Jacob Burns and Steffen Kraemer presented their work for Forensic Architecture about drone strikes in Afghanistan that keep making civilian casualties without the world responding. See: www.forensic-architecture.org/case/drone-strikes/ (accessed 4 December 2017).

14  See Brian Eno's argument in his essay 'The big here and long now' (Eno, 2004).

15  Ironically, urban theorising moves away from place-based container views towards a networked epistemology of cityness (e.g. Brenner and Schmid, 2015), and media theory after the 'spatial turn' becomes increasingly location-specific and situational. Yet common ground remains shaky.

## References

Adam, B. (1990) *Time and Social Theory*. Philadelphia, Pennsylvania: Temple University Press.

Adam, B. (2008) Of timespaces, futurescapes and timeprints. [Online] Available at: www.cardiff.ac.uk/socsi/futures/conf_ba_lueneberg170608.pdf (accessed 1 August 2016).

Ariely, D. (2008) *Predictably Irrational: The Hidden Forces that Shape Our Decisions*. New York: Harper Collins Publishers.

Arvidsson, A. and Colleoni, E. (2012) Value in informational capitalism and on the internet. *The Information Society*, 28(3): pp. 135–150.

Balestrini, M. (2013) In favour of a multiplied self. Can empathy lead to personal behaviour change? CHI'13 Conference, Paris: France, April 27–May 2 2013. [Online] Available at: www.personalinformatics.org/docs/chi2013/balestrini.pdf (accessed 25 August 2016).

Batty, M. (2015) A perspective on city dashboards. *Regional Studies, Regional Science*, 2(1): pp. 29–32.

Bleecker, J. and Nova, N. (2009) 'A synchronicity: Design fictions for asynchronous urban computing'. In: Khan, O., Scholz, T. and Shepard, M. (eds) *Situated Technologies Pamphlet Series*. New York: The Architectural League of New York. [Online] Available at: www.situatedtechnologies.net/files/ST5–A_synchronicity.pdf (accessed 12 December 2007).

Bolter, J. D. and Grusin, R. (2000) *Remediation: Understanding New Media*. Cambridge, Massachusetts: MIT Press.

Bouman, M. (2013) 'Move along folks, just move along, there's nothing to see: Transience, televisuality and the paradox of anamorphosis'. In: de Valck, M. and Teurlings, J. (eds) *After the Break: Television Theory Today*. Amsterdam: Amsterdam University Press, pp. 161–177.

Brenner, N. and Schmid, C. (2015) Towards a new epistemology of the urban? *City*, 19(2–3): pp. 151–182.

Buser, M. (2014) Thinking through non-representational and affective atmospheres in planning theory and practice. *Planning Theory*, 13(3): pp. 227–243.

Callon, M. and Law, J. (2003) On qualculation, agency and otherness. [Online] Available at: www.lancaster.ac.uk/fass/resources/sociology-online-papers/papers/callon-law-qualculation-agency-otherness.pdf (accessed 1 August 2016).

Ciuccarelli, P., Lupi, G. and Simeone, L. (2014) *Visualising the Data City: Social Media as a Source of Knowledge for Urban Planning and Management*. Dordrecht: Springer.

Crandall, J. (2010) The geospatialisation of calculative operations: Tracking, sensing and megacities. *Theory, Culture and Society*, 27(6): pp. 68–90.

Crang, M. and Graham, S. (2007) Sentient cities: Ambient intelligence and the politics of urban space. *Information, Communication and Society*, 10(6): pp. 789–817.

de Lange, M. (2010) Moving circles: Mobile media and playful identities. Unpublished thesis (PhD), Erasmus University Rotterdam, Rotterdam.

de Lange, M. (2013) 'The smart city you love to hate: Exploring the role of affect in hybrid urbanism'. In: Charitos, D., Theona, I., Dragona, D. and Rizopoulos, H. (eds) *The Hybrid City II: Subtle Revolutions Proceedings*, Athens, Greece, 23–25 May.

de Lange, M. and de Waal, M. (2012) Social cities of tomorrow: Conference text. [Online] Available at: www.socialcitiesoftomorrow.nl/background (accessed 1 August 2016).

de Lange, M. and de Waal, M. (2013) Owning the city: New media and citizen engagement in urban design. *Media and the City*, 18(11). [Online] Available at: http://firstmonday.org/ojs/index.php/fm/article/view/4954/3786 (accessed 1 August 2016).

de Waal, M. and de Lange, M. (2014) 'Klik. Like! Share: Hoe digitale media de publieke ruimte veranderen'. The Mobile City/Publieke Ruimte – Publieke Zaak. [Online] Available at: http://themobilecity.nl/wp-content/uploads/2014/12/The_Mobile_City_PRPZ_KlikLikeShare.pdf (accessed 1 August 2016).

Diakopoulos, N. (2014) Algorithmic accountability reporting: On the investigation of black boxes. The Tow Foundation and the John S. and James L. Knight Foundation: Colombia Journalism School. [Online] Available at: http://towcenter.org/wp-content/uploads/2014/02/78524_Tow-Center-Report-WEB-1.pdf (accessed 1 August 2016).

Eno, B. (2004) The big here and long now. *Digitalsouls. com*, 27.

Flowers, M. (2013) 'Beyond open data: The data-driven city'. In: Goldstein, B. and Dyson, L. (eds) *Beyond Transparency: Open Data and the Future of Civic Innovation*. San Francisco, California: Code for America Press, pp. 185–198.

Gabrys, J. (2014) Programming environments: environmentality and citizen sensing in the smart city. *Environment and Planning D: Society and Space*, 32(1): pp. 30–48.

Giddens, A. (1991) *Modernity and Self-identity: Self and Society in the Late Modern Age*. Stanford, California: Stanford University Press.

Gladwell, M. (2005) *Blink: The Power of Thinking Without Thinking*. New York: Little, Brown and Company.

Greenfield, A. and Shepard, M. (2007) 'Urban computing and its discontents'. In: Khan, O., Scholz, T. and Shepard, M. (eds) *Situated Technologies Pamphlet Series*. New York: The Architectural League of New York. [Online] Available at: www.situatedtechnologies.net (accessed 12 December 2007).

Hannerz, U. (1980) *Exploring the City: Inquiries Toward an Urban Anthropology*. New York: Columbia University Press.

Hind, S. and Gekker, A. (2014) 'Outsmarting traffic, together': Driving as social navigation. *Exchanges: The Warwick Research Journal*, 1(2): pp. 165–180.

Holden, M. and Moreno Pires, S. (2015) Commentary on Rob Kitchin et al.'s 'Knowing and governing cities through urban indicators, city benchmarking, and real-time dashboards'. *Regional Studies, Regional Science*, 2(1): pp. 33–38.

Kitchin, R. (2013) *The Real-Time City? Big Data and Smart Urbanism*. The 'Smart Urbanism: Utopian Vision or False Dawn' workshop at the University of Durham, 20–21 June 2013. [Online] Available at: https://papers.ssrn.com/sol3/papers.cfm?abstract_id=2289141 (accessed 1 August 2016).

Kitchin, R. (2014) *The Data Revolution: Big Data, Open Data, Data Infrastructures and Their Consequences*. London and Thousand Oaks, California: Sage Publications.

Kitchin, R., Lauriault, T. P. and McArdle, G. (2015a) Knowing and governing cities through urban indicators, city benchmarking and real-time dashboards. *Regional Studies, Regional Science*, 2(1): pp. 6–28.

Kitchin, R., Lauriault, T. P. and McArdle, G. (2015b) Urban indicators and dashboards: epistemology, contradictions and power/knowledge. *Regional Studies, Regional Science*, 2(1): pp. 43–45.

Klauser, F., Paasche, T. and Söderström, O. (2014) Michel Foucault and the smart city: Power dynamics inherent in contemporary governing through code. *Environment and Planning D: Society and Space*, 32(5): pp. 869–885.

Levitt, S. D. and Dubner, S. J. (2005) *Freakonomics: A Rogue Economist Explores the Hidden Side of Everything*. London: Allen Lane.

Lindner, C. (2013) Smart cities and slowness. *Urban Pamphleteer*, 1: pp. 14–16.

Lynch, K. (1960) *The Image of the City*. Publications of the Joint Center for Urban Studies. Cambridge, Massachusetts: The Massachusetts Institute of Technology Press.

Mattern, S. (2015) Mission control: A history of the urban dashboard. *Places Journal*, March. [Online] Available at: https://placesjournal.org/article/mission-control-a-history-of-the-urban-dashboard/ (accessed 1 August 2016).

Medina, E. (2006) Designing freedom, regulating a nation: Socialist cbernetics in Allende's Chile. *Journal of Latin American Studies*, 38: pp. 571–606.
Morozov, E. (2013) *To Save Everything, Click Here: The Folly of Technological Solutionism*. New York: Public Affairs.
Morozov, E. (2014) The planning machine: Project cybersyn and the origins of the big data nation. *The New Yorker*, 13 October. [Online] Available at: www.newyorker.com/magazine/2014/10/13/planning-machine (accessed 1 August 2016).
Perez, T. and Rushing, R. (2007) *The Citistat Model: How Data-Driven Government Can Increase Efficiency and Effectiveness*. [Online] Available at:https://cdn.americanprogress.org/wp-content/uploads/issues/2007/04/pdf/citistat_report.pdf (accessed 1 August 2016).
Peters, J. D. (1999) *Speaking into the Air: A History of the Idea of Communication*. Chicago, Illinois: University of Chicago Press.
Ricoeur, P. (1992) *Oneself as Another*. Translated by K. Blamey. Chicago, Illinois: University of Chicago Press.
Rodgers, S., Barnett, C, and Cochrane, A. (2014) Media practices and urban politics: Conceptualising the powers of the media-urban nexus. *Environment and Planning D: Society and Space*, 32(6): pp. 1054–1070.
Rozin, P., Haddad, B., Nemeroff, C. and Slovic, P. (2015) Psychological aspects of the rejection of recycled water: Contamination, purification and disgust. *Judgment and Decision Making*, 10(1): pp. 50–63.
Thrift, N. J. (2008) *Non-Representational Theory: Space, Politics, Affect*. International Library of Sociology. New York: Routledge.
Townsend, A. (2000) Life in the real-time city: Mobile telephones and urban metabolism. *Journal of Urban Technology*, 7(2): pp. 85–104.
Tuters, M. and de Lange, M. (2013) 'Executable urbanisms: Messing with Ubicomp's singular future'. In: Buschauer, R. and Willis, K. S. (eds) *Locative Media: Multidisciplinary Perspectives on Media and Locality*. Bielefeld: Transcript, pp. 49–70.
Van der Spek, S., Van Schaick, J., De Bois, P. and De Haan, R. (2009) Sensing human activity: GPS tracking. *Sensors*, 9(4): pp. 3033–3055.
Wakefield, S. and Braun, B. (2014) Governing the resilient city. *Environment and Planning D: Society and Space*, 32(1): pp. 4–11.
Walker, R. (2011) The lie of the land: Mark Monmonier on maps, technology and social change. *Visual Studies*, 26(1): pp. 62–70.
Williams, A. and Srnicek, N. (2013) Accelerate manifesto for an accelerationist politics. [Online] Available at: http://criticallegalthinking.com/2013/05/14/accelerate-manifesto-for-an-accelerationist-politics/ (accessed 1 August 2016).
Wilson, M. W. (2015) Flashing lights in the quantified self-city-nation. *Regional Studies, Regional Science*, 2(1): pp. 39–42.
Zeidner, M., Matthews, G. and Roberts, R. D. (2009) *What We Know About Emotional Intelligence: How It Affects Learning, Work, Relationships, and Our Mental Health*. Cambridge, Massachusetts: The Massachusetts Institute of Technology Press.

# 12

## Conclusion: back to the future

*Alex Gekker, Sam Hind, Sybille Lammes, Chris Perkins and Clancy Wilmott*

The chapters in this book have emerged and changed in an ongoing process. You are reading something that is apparently fixed, with an endpoint in terms of its production. But tracing the emerging ideas back to an origin is a much more challenging task; each chapter – and indeed this conclusion – emerged gradually and in a nonlinear fashion. Events can be planned or random, they may be fast or slow and digital tools may allow us to accelerate or impede this process. New digital technologies also afforded us the chance to enact specific rhythms in the writing process. We worked quickly on this book as editors, by organising a series of book-sprints in remote areas in the United Kingdom and the Netherlands, as well as across digital and virtual time-spaces. During this process, we collaborated on editing chapters simultaneously and jointly – composing the conclusion and introduction in shared Google Docs, watching and over-writing each other's work. How we, during a shared yet sometimes haphazard time frame of days, hours or moments, worked on this book serves as a useful allusion to how temporality has taken on new meanings since the advent of digital mapping.

Much like the digital mapping practices we have discussed in this book, time proved to be tricky, asynchronous, serendipitous, sticky and ephemeral, thanks to – or despite – digital technologies. Someone was sent out of the room where she Skyped, someone else had to walk a dog, was hungry and had to grab lunch, forgot about time differences, spilled a drink over a computer, or failed to find the right document. Every time we faced an interruption, we had to try to realign our times so as to recover glimpses of immediacy. The goal was for the palimpsest of an emerging document to become a conceptual landscape, resonating with the sometimes false promises of immediacy that digital mapping so often seems to deliver.

## Conclusion: back to the future

Based in different countries and universities, we came together in shared spaces to deliver an apparently fixed outcome; akin to how crowd-sourced mapping emerges from differently situated participants. The process of making this book is the result of an effort to synchronise time over different places. It shows how the digital can invite us to be in a different place at the same time, tracking each other's emerging paths from a distance. Yet it also shows how time and technologies can subvert our expectations, lead to unexpected results and otherwise fail. For example, Skype connections on mobile phones picked up background noises which obscured the actual voice of the speaker, and electronically edited documents escaped attention due to the temporal simultaneity of this medium. So the words may be about time but they also offer a glimpse into aspects of time as they emerge from an ongoing process of writing. This process of writing stands in many ways as analogous to mapping, and towards the end of a process, concluding chapters traditionally reflect on what has been delivered and on what the future might entail. Whereas a digital map is never complete (Kitchin and Dodge, 2007), this book is ending. So, looking back and forward – much like a Trip Advisor restaurant review reflecting on a meal, but also inviting others to experience it – we argue that it is about time!

Chapters in this book have explored conceptual frameworks for understanding the rhythms, ephemerality, transformativity, futurity and immediacy of digital mapping. They therefore demonstrate why these temporal concepts are needed to start to make sense of new cartographic forms emerging after the advent of the digital. Be it in art, daily life or planning, data production and consumption continue to accelerate apace, and new affordances emerge for engaging with maps accordingly. Mobility further complicates this unfolding of contemporary cartographies and brings with it new conceptions of temporality, as the map itself and the user are not only on the move, but also reciprocally transform each other through time, processes that are often difficult to grasp.

This book has offered some tools to address these changes. Of course, when temporalities become privileged, this has a reciprocal effect on how we view space. Previously hidden temporal conceptualisations may be needed to understand digital mapping practices, but they do not exist independent of spatiality. The concepts explored in this book communicate ideas, even when they are being used as an approach, and in turn impact on mapping practices. In this conclusion, we take stock and look back at what our temporal foci has done for time-space and its imagining as a process, highlighting questions that have emerged during this process.

This processual quality of temporality is important, but many other ways of approaching time might be deployed. We started our consideration of temporality in the introduction by deploying Barbara Adam's (2008) sevenfold classification of time as a means of discussing different philosophical approaches to them. Adam highlights seven features of time: time frame; temporality; timing; tempo; duration; sequence and modalities (for an explanation of their significance see chapter 1 this volume). By examining each section of this collected volume in relation to its emerging internal consistency with Adam's typology, we highlight how different authors foreground different aspects of temporality. In doing so, we suggest aspects that have perhaps been underanalysed and where future research might profitably focus.

## Ephemerality/mobility

Against the grand scales of history, ephemerality and mobility offer useful entry points into a nuanced discussion about everyday temporality, precisely because of the complex questions they pose for mapping. On the one hand, fleeting encounters and irretrievable processes engage mapping as an iterative and temporary practice, that works in both performative and representational ways, and which builds and dissolves according to heterogeneous modalities. On the other hand, mapping 'products' frequently retain a longevity across multiple and ephemeral platforms – such as OpenStreetMap, Google Street View or McLean's smell maps. The ways in which each platform presents a version of temporality is encased in their own unique time frames. Questions arise in this disjunction between the ephemeral (or that which doesn't last), presented by the analysis of encounter by Gerlach, McLean and Abend, and the lingering traces, data sweat (Gregg, 2015) – or data fumes (Thatcher, 2014) – that endure beyond the moment. A key issue is how do we resolve the question of scale when considering temporality? For space and place, we already have some answers – Massey (1991: 24) suggests it is imperative to reconsider places, not as a bounded and defined, but as spaces with histories. This is particularly salient at a time 'when things are speeding up' due to time-space compression and when the global and the local become ever more intimately intertwined. It is a time when 'history itself is imagined as the product of layer-upon-layer of different sets of linkages, both local and to the wider world' (Massey, 1991: 29). But if we were to invert that assertion, as we have tried to do throughout this volume, to focus not on spatio-temporality, but on tempo-spatiality, then a different set of questions emerge.

First, a temporal focus draws attention to multifaceted sets of scale, which appear to braid into different temporal durations at social and personal levels, in an age when time appears to accelerate exponentially. Walter Benjamin once described this as *Erfahrung* and *Erlebnis* (the temporal distinctiveness of near and far experience and memory) (see Elsaesser, 2009), and the digital mapping cases discussed in this section also evoke Bergsonian notions of *durée* and Adam's (2008) notion of duration. Gerlach in particular highlights the implications of an approach focusing on these durative, as against chronological or sequential, issues. We can also see this process re-enacted in many examples of memory work, deployed close to hand, but also using digital connections to call into play remote experiences. Tracks and points are employed by the Intel Corporation as a system for data retention in computing, delivering a shared metric for everyone, generalised through GPS models and readily available beneath smartphone interfaces. But this shared cartographic reason (Olsson, 2007) is deployed by individuals whose experiences and memories of experiences are locally enacted and of the moment. As such, a mobile and personal experience of temporality reveals tensions in digital mapping (Wilmott, 2016). Time is global, but that registers and regulates against temporality, which is lived and experienced. These tensions are further explored in this section. Gerlach waits for his GPS device to arrive, and when it does, the device records and reinforces personal narratives at thirty-second intervals. Similarly, McLean's plural mapping processes makes links between people, smells and space, and the end product, a map, which at once challenges what has traditionally been perceived as being cartographic, while at the same time capturing a series of ephemeral moments, a flow, by way of traces on paper. Our third case from Abend recounts the glitches he encounters which fragment and disrupt the uniformity of time ambitiously presented by Google Street View. Once again, the near and the far merge in a technological process. There is a rich potential for research that focuses upon these scalar tensions, and its inherently social focus lies beyond Adam's (2008) classification.

A second set of questions arise because the ephemeral cannot be disentangled from the epochal as easily as we imagine. Digital mapping builds on cartographic traditions that are thoroughly epochal traditions which are structured and scaffolded to scale from the very local and immediate, to the very global and distant. The glimpse of a suburban street-in-action, or the everyday vernacular mappings of a local OpenStreetMap group, or the recording of transient sensation of urban smells, are juxtaposed to massive collections of data accrued through millions of these moments. This scaling delivers the power of maps (Wood, 1992). But it is no longer through the representational fixity of pen on paper, or the temporal selection of the image, and no longer by the aesthetic or artistic practice of cartographers and groups of surveyors employed by nation-states to fix and frame their territories. Instead, the presence of the Google car makes these links possible, and formerly clear distinctions between users and producers blur in the world of crowd-sourcing and the 'gig' economy. A very different digital assemblage comes together to make these links, which makes for changed affordances and changing ways in which the temporal is captured. There are further discussions to be had here about these interrelations between heterogeneous temporalities, the relationship between mapping processes and mapping products, and the arbitrary designation of timescales into frames – seconds, minutes, hours, days, weeks, or months – through which we understand time. Once again, the neat classification from Adam breaks down and once again, research is needed to explore the relations that emerge.

A third and crucial question that emerges from this section relates to the temporal layering that is implicit in most digital mapping (see Verhoeff, 2012). Massey (1991) asks how space enrols time, but how can we rethink a temporality which pulls in layers of spaces and places? She describes a historical space, which is the product of interlinked layers of historical moments and events, but layered digital mapping enrolling space into time complicates outcomes. What does it mean to reject the clarity of borders between history and experience, between the global stories that build empires and nations into perpetuity (even if only in the archive), and those that build memories and everyday localities? The precise role that digital mapping might play in this process remains to be charted in detail, beyond the cases suggested in our chapters.

So a focus on ephemerality and mobilities suggests research agendas that are rather more performative, transient and hybrid than might be imagined in the at times static implications of Barbara Adam's (2008) typology.

## Stitching memories

In this section of the book, we drew attention to how the digital has foregrounded an accelerated process of combining and recombining cartographical data, and how tempo has encouraged us to grapple for new ways to remember and capture places and trajectories when it all seems to slip to our fingers so easily. Our authors ask what practices might exist to counter the ephemeralities and mobilities that digital mapping has brought with it, which were discussed in the first section of the book. As we are dealing with assemblages of digital technologies, their constant shifting employment can lead to an impression of time 'flying', and of an inevitable mutability. But the very same technologies can also be aligned and mastered to re-appropriate experiences and to stitch memories together in novel ways. This section was about reflection and looking back, to track what has happened on the map and create 'bricolages' with their own tempos and histories.

Authors in this part of the book discussed daily and creative practices that actually address some of the questions that the first section of the book left us with, and especially the role that digital mapping might play in building memories by creating new temporal tapestries. As this section is concerned with the capturing and recombining of different moments and memories, the emphasis shifts from Adam's (2008) categories of duration to tempo; the speed, but also the stillness, that is called into play when slivers of time are recombined, and how such capturing processes are visualised as staccato cartographical fabrics. This stitching inevitably only delivers an edited version of the past: events and experiences can be edited and recombined, forming new patterns of memories, but also frequently hide the editorial processes that were employed to make the stitch-work!

The first two chapters are mostly concerned with artistic interventions that use digital tools to map past time, and find new ways of producing stories through them, much in the same vein as the smell maps of Kate McLean seek to make narrative accounts of smellscapes. However, unlike McLean's work, they empathise the sutures instead of temporal flow. Rachel Wells' chapter about the temporal maps created by visual artist Wolfgang Weileder discusses the reordering of temporalities. In his *atlas-project*, he stacks slices of photographic renderings of the same place on top of each other to create a visual rendering of different moments stitched together through techniques of digital editing. Likewise, MacDonald shows how art projects can capture mapping processes by recounting mapping moments in the past and bringing them together in new narratives, thus showing how remnants of past mapping processes can form new productive relations for remembering place. Such attempts are not exclusive to the realm of artistic practice though, but can also be found in daily life, as Hanchard shows in his chapter. Advocating a practice theory perspective on quotidian mapping, he shows an intricate patchwork of stories through which people remember and reflect on quotidian mapping practices. So, by stressing the temporal dimension of cartographical stitching of memories and stories, we have showed here some counter-strategies to the ephemeral quality of digital mapping.

The first question that arises from these chapters concerns how successful these strategies might be for facilitating resistance – when technologies encourage precarity, transience and dynamism, or in a wider context how stitching might gain cultural currency. It is notable that two of the three cases described here are artistic interventions designed to question and critique, and thus not necessarily indicative of the temporalities in mass-consumption digital cartography. Stitching may well be a problematic strategy for the mainstream in an age when tempo has shifted to discourage fixing, in favour of flow. Recent research focusing upon narrative cartography (Caquard, 2013) suggests some of the ways in which memories are increasingly being woven into personal cartographies.

We suggest that this stand might usefully be developed and extended to new fields beyond the artistic.

A second question that emerges from the strategies described here concerns the affordances that they facilitate (Gibson, 1977; Greeno, 1994). To work, mapping needs to be deployed, read and enacted. But very little research to date has focused on interactionist ecological investigation of how a fixing together of temporal frames might actually be read. Stitched and recombined bricolages, like the artworks produced by Wolfgang Weileder and described by Rachel Wells in chapter 5 of this volume, are challenging to interpret, but paradoxically have to be read if they are to gain cultural currency. We need to find out how memory maps do their work.

A third question arising from the stitching together concerns the implication for spatiality. We would argue that it tends to focus attention on the particular and the unique, and onto places as against spaces. Places, in these stitchings, become holders for memory, moments or rhythms, that might be recombined in a reflective story of mapping with particular and unique resonances. To look at these kinds of practices through a temporal lens does not so much obliterate spatiality, but instead alters the analysis beyond modern conceptions, shifting to a more hybrid understanding of tempo-spatial translations. There is nothing inevitable about flow, and rich place-based analyses are needed to explore how the politics of placeholding emerges in a transient age.

So the capacities of digital mapping to stitch, as against to split, also demand attention. Adam's (2008) notions of tempo, sequence, time frame and rhythm are embodied in critical questions of *how* stitching can take place in an age when technological context tends to privilege mutability.

## (In)formalising

This section exemplified the various ways in which digital maps shape, predict and invent time. In the introduction to this book, we suggested that technological cartographic projects have either sought to resist inherent temporal instability (such as weather or disease maps), or else sought to build and expand upon it (e.g. various social media-based reaction maps). A tension thus exists between the formalisation of time, and its informal destabilisation. Michiel de Lange's chapter complicates this unstable notion of temporality. Discussing the problematics of the smart city dashboard's 'real-time', he builds on sociologist Barbara Adam's (2008) work to further disassemble the idea of temporal instability. He argues for an asynchronous approach to the city that 'highlights latency, recurrence, deferred understanding and imperfections of the mediating process' in mapping the smart city (de Lange, this volume). The other chapters in this section also extend and call into question temporal logics, in and beyond the smart city. Sutherland's work challenges – through the notion of flows – the tempo of digital maps. Cate Turk's work on foam ontologies challenges established notions of sequences. Tuur Driesser reverses the temporal modalities present in a PathoMap. Each of them goes beyond Adam's classification to explore processes and relations, and each focuses on different aspects of this (in)formalisation.

As Sutherland has noted in his chapter, the mapping of flows is a well-established domain and has come to be associated with more recent global, digital manifestations of capital circulation. The obsession with rendering the world as flow-full (mirroring similar obsessions for the real-time in de Lange's piece, and prediction/prevention in Driesser's), therefore, has led to a curious abandoning of anything fixed, grounded or otherwise stable – despite the continuing presence of infrastructures that facilitate the various flows of capital around the world. Recently, there has been a growing academic interest in the 'logistics' of these flows. That is to say, the systems, architectures and processes that allow for the near-constant movement of goods and people around the world. Ned Rossiter's (2015) work on the global logistics industry and its intertwining with new digital infrastructures and software architectures is particularly relevant in this context.

So, a first key question lies with the materiality of time. If triumphalist narratives proclaiming the free-floating nature of global life are to be avoided, then these nascent interventions are critical. Moreover, things do not move on their own, but are reliant on huge material investments in labour power as well as fixed capital. All too frequently, these aspects are ignored or taken for granted. There is, of course, a temporal specificity to such operations and a requirement that particular activities are carried out in advance of, or just in time for, their use. Think, for example, of global logistics firms such as FedEx, whose current advertising strapline is 'The World on Time'.[1] Needless to say, there is much more to be done on this logistical time that not only takes account of how this logistical logic plays out, but indeed where it plays out. How these logistical operations are governed, streamlined, standardised and audited by various integrated modules and systems being put to work the world over, literally makes a difference. As is thus evident, there are important material and social concerns beyond Adam's framework, which demand attention.

A second series of questions concerns the conceptual frameworks that might usefully underpin research. These formalised structures themselves strongly influence interpretations of digital mapping. There have been a multitude of academic appraisals of flow-thinking and the mapping of flows, from nodes and across networks (see for example Latour, 1986; Deleuze and Guattari, 1988; Castells, 1991). While the taking-up of new conceptual frameworks should always be done with caution, so as to avoid blindly overlaying an otherwise ill-fitting set of ideas on top of already existing practices and processes, Sloterdijk's (1998; 1999; 2004) work on foams presents some interesting opportunities. Indeed, Turk's question of whether 'freezing' phenomena is necessary in order to analyse them – say, in relation to the dynamism of digital mapping interfaces – is richly answered by deploying Sloterdijk. Understanding the way in which various elements combine within a more general architecture, say, in relation to humanitarian projects, is critical. In deploying bubbles and foams, Turk makes a much-needed abstract turn. She implicitly attacks naïve behaviouralist or technicist approaches, frequently seen in new empiricist claims of big data research that privilege technologies (see Kitchin, 2014). But the emphasis on bubbles and foams also critiques Deleuzian or Latourian thought that presupposes a world of networked flow, or a world of nodes and translations. While other theoretical framings stress the 'unfolding' or 'enveloping' properties of time that speak only to a kind of 2D world, Sloterdijk's bubbles and foams point towards a 3D world in which the 'globular' or the 'atmospheric' take centre-stage. The question for future researchers is how conceptual vocabularies might be developed, through which the temporal and the cartographic can be spoken of as being in concert?

A third key question emerges in Tuur Driesser's contribution. How might the future be foreclosed or anticipated in new mapping projects that seek to anticipate emergent risks and situations? While Driesser adopts a novel 'object-oriented' approach – relative to, but not entirely the same as, ANT – there are other nascent avenues that must be considered in light of the ominous suggestion that the contingency of future actions is being shut down (see for example Anderson, 2010). Unlike the material and conceptual provocations above, Driesser probes an intensely political force to speculate on what might be lost – socially, culturally, temporally and cartographically – in light of new biopolitical developments that arguably threaten rather than enhance current ways of living. In this chapter, temporality becomes a new battleground over which political forces exert their will on bodies, objects and relations. Although more optimistic contributions stress that these speculative thoughts are but hopeful 'visions' rather than material realities, the fact remains that they drive a host of actually existing projects and initiatives, many of which are subsumed at present under the 'big data' umbrella embodying a 'capture all' mentality and desiring an 'anticipate it all' future. Perhaps more than anything else, Driesser's intervention seeks to establish the coordinates for this new future-oriented world and as a clarion call for a political consideration of temporalities.

The authors of (in)formalising have focused not only on the inherent differences between the various temporalities enacted in digital maps, but – and perhaps more crucially – on the troubles of comparing and understanding the nature of such temporalities in relation to one another. The structures beyond time powerfully affect material, conceptual and anticipatory logics enrolled into temporality. Adam's typology significantly underplays these organisational, infrastructural and formal tropes.

## It's about time

This book has taken the reader on a journey through time, revealing ways in which digital mapping can be approached using temporal lenses, from different theoretical positions or methodological ways of doing. May and Thrift (2001) chose to focus on time-space as a series of concepts impacting on wider issues relating to practice across geography, whereas here we emphasise the very particular case of technological change, and the different ways in which a digital remaking of mapping might relate to temporality. The two decades since their intervention have seen significant social, economic and political shifts towards increasingly digital life, with consequent and profound implications for time and temporality. Authors from different interdisciplinary backgrounds have shown what happens when temporality is privileged over spatiality in the analysis of digital mapping. That does not mean that the book sheds the notion of spatiality altogether, but rather that the contributions have mainly worked from tempo-spatiality instead of spatio-temporality.

We have suggested that key research agendas might usefully be progressed by further exploration of ephemeralities and mobilities, stitching and (in)formalising aspects of temporality. Sometimes these suggestions are practical; attending to the mechanisms and affordances of how stitching might occur for example. But changing times also demand more conceptual innovation, such as consideration of how epistemology and formal structures of knowledge impact researching temporalities of digital mapping. The theories and methodologies that emerged through these very different interventions can help us to refine and extend our understanding of temporality in relation to digital mapping and digital culture in general. It is about time, because, as the contributions of this book show, digital mapping interfaces are making new temporal experiences that hitherto have not been sufficiently addressed by approaches to mapping that emphasise spatiality of the medium. It is about time, because mapping has begun to be re-thought in productive temporal terms.

## Note

1 See: http://i1108.photobucket.com/albums/h420/Dr3am3r1011/IMG_2719.jpg (accessed 4 December 2017).

## References

Adam, B. (2008) Of timespaces, futurescapes and timeprints. [Online] Available at: www.cardiff.ac.uk/socsi/futures/conf_ba_lueneberg170608.pdf (accessed 1 August 2016).
Anderson, B. (2010) Preemption, precaution, preparedness: Anticipatory action and future geographies. *Progress in Human Geography*, 34(6): pp. 777–798.
Caquard, S. (2013) Cartography: Mapping narrative cartography. *Progress in Human Geography*, 37(1): pp. 135–144.
Castells, M. (1991) *The Informational City: Information Technology, Economic Restructuring, and the Urban-Regional Process*. Oxford: Blackwell.
Deleuze, G. and Guattari, F. (1988) *A Thousand Plateaus: Capitalism and Schizophrenia*. London: Bloomsbury Publishing.
Elsaesser, T. (2009) Between Erlebnis and Erfahrung: Cinema experience with Benjamin. *Paragraph*, 32(3): pp. 292–312.
Gibson, J. J. (1977). 'The theory of affordances'. In: Shaw, R. and Bransford, J. (eds) *Perceiving, Acting, and Knowing: Toward an Ecological Psychology*. Hillsdale, New Jersey: Erlbaum, pp. 67–82.
Greeno, J. G. (1994) Gibson's affordances. *Psychological Review*, 101(2): pp. 336–342.
Gregg, M. (2015) Inside the data spectacle. *Television and New Media*, 16(1): pp. 37–51.
Kitchin, R. (2014) Big data, new epistemologies and paradigm shifts. *Big Data and Society*. [Online] Available at: http://bds.sagepub.com/content/spbds/1/1/2053951714528481.full.pdf (accessed 31 August 2016).
Kitchin, R. and Dodge, M. (2007) Rethinking maps. *Progress in Human Geography*, 31(3): pp. 331–344.
Latour, B. (1986) 'Visualization and cognition: Thinking with eyes and hands'. In: Kuklick, H. (ed.) *Knowledge and Society: Studies in the Sociology of Culture Past and Present Volume 6*. New York: Jai Press, pp. 1–40.
Massey, D. (1991) A global sense of place. *Marxism Today*, June: pp. 24–29.
May, J. and Thrift, N. (eds) (2001) *Timespace: Geographies of Modernity*. London: Routledge.
Olsson, G. (2007) *Abysmal: A Critique of Cartographic Reason*. Chicago, Illinois: University of Chicago Press.
Rossiter, N. (2015) Coded vanilla: Logistical media and the determination of action. *South Atlantic Quarterly*, 114(1): pp. 135–152.
Sloterdijk, P. (1998) *Sphären I – Blasen*. Frankfurt: Suhrkamp.
Sloterdijk, P. (1999) *Sphären II – Globen*. Frankfurt: Suhrkamp.
Sloterdijk, P. (2004) *Sphären III – Schäume*. Frankfurt: Suhrkamp.

Thatcher, J. (2014) Living on fumes: Digital footprints, data fumes, and the limitations of spatial big data. *International Journal of Communication*, 8: pp. 1765–1783.
Verhoeff, N. (2012) *Mobile Screens: The Visual Regime of Navigation*. Amsterdam: Amsterdam University Press.
Wilmott, C. (2016) Small moments, big data: Mobile mapping in everyday life. *Big Data and Society*, 3(2). https://doi.org/10.1177/2053951716661364.
Wood, D. (1992) *The Power of Maps*. New York: Guilford Press.

# Index

Abend, P. 15, 91–111
acceleration 125, 245
actor network theory 7, 158, 166, 168, 169, 201–202
Adam, B. 5–6, 242–248, 257–258
affect 232–233
affordances 262
Agamben, G. 231
agency/structure dualism 159
algorithmic framing 104–106
*Amsterdam RealTime* 139–140
anamorphosis 249
anchoring socio-temporal practices 165–168
Anderson, B. 229
animation 57–58, 95–100, 198–200
anticipatory logic 17, 18, 229, 232
Appadurai, A. 183–184
Application Programming Interface 156
apps 67–69, 167
artistic practice 50–90, 115–136, 138–153, 261
*Aspen Movie Map* 95–96
assemblage 202, 203, 205, 207, 224, 231, 259–260
asynchronicity 19, 238–251
    politics 249–250
    representation 248–249
Asynchronous Javascript and XML 156
atemporality *see* timelessness
*atlas-project* 16, 119–121
augmented reality 148–149
authorship 161

*Back to the Future* 2–3, 256–267
Bakhtin, M. 8
balloons 139
Barker, T. 200, 207–208

Barthes, R. 123
Benjamin, W. 93, 108, 121–123, 124, 129, 135
Bennett, J. 226, 234
Bergson, H. 6, 28, 34, 40–41, 46, 101–102, 106–107, 108
biopower 7
bodily experience 37–38
Bourdieu, P. 160–161, 166, 167
bricolage 214, 260
bubbles 204–205

camera 116, 117, 118, 120, 127, 128, 133
*Camera* 132–134
cars and smell 84
cartographic reason 28, 36, 45, 177, 259
cartography
    colonial history 45, 177–178
    critical 155–156, 225–226, 234–235
    existential crisis 30
    history of 156
    mediatisation 97–98
    narrative 261
    object-oriented 225–227
    post representational 226–227
    theory 155–159, 225–226
Castells, M. 142, 148, 185
Center for Urban Science and Progress New York University 223, 227
center of calculation 92
*Chronicles of Eusebius* 50
chronotope 8
cinematic film 94, 95
*A Clear Day* 145
clock time 142
CloudMade 33
colour 86–87

# Index

conceptual frameworks 263–264
convergence 95, 96
Coordinated Universal Time 142
coordinates 31–32, 117, 126, 133, 135
copyright 37
*Crank* 96–97
crisis mapping 18, 208–219
   project histories 217
crowd sourced maps 27, 30, 34, 215, 216, 217
crystal image 108
cultures of map use 160, 161
Cybersyn Control Center 239–240

Daguerreotype 136
dashboards *see* urban dashboards
de Certeau, M. 58, 144
de Lange, M. 18–19, 238–255
Deleuze, G. 6–7, 94–95, 106, 207–208
Deliveroo 12
Derrida, J. 7, 142
Digital Earth 11, 103
digital maps
   affordances 163
   culture of 161
   material artefacts 7, 18, 161
   memory 163–164
   technological innovation 160
   use 157, 198–200
digital photography 115–137
dimensionality 264
Dionysian 126, 127, 128–129
DNA sequencing 224
Driesser, T. 18, 223–237
duration 6, 28, 69–80, 106–107, 245–246, 259
dynamic maps 198–200, 208–219

Ebsdorf map 101
emotion 74
Engel, C. 104–105
ephemerality 14–15, 25–112, 55–57, 218, 258–260
epochal timeframe 259
Erfahrung 258–259
Erlebnis 258–259
errors 13, 38
event 6

everyday geography 28–29, 155–172
executable urbanism 249

failure 31
*Feedlots* 146–148
*FieldWorks* 144
fieldwork traces 29–32
film 93, 94–98, 108
flow 176
   given-ness 186–193
   social sciences 183–186
flow chart 180
flow maps 176–183
fluid topology 229–231
foams 204–205
forecasts 85–86
framing 98, 109
free smelling 83
freezing time 79–80, 165–168
Fujihata, M. 144
*Fukashima Online Issues Map* 203
future present 247
futurity 3, 7, 66–67, 85, 264

Galloway, A. 189
Gekker, A. 1–23, 256–267
gentrification 77–78
geobrowsing 100–103, 104
geographical information system (GIS) 156
geomedia 91–111, 138–153
Gerlach, J. 14–15, 27–49
gig economy 12, 259
GIS wars 156
glitches 31, 106, 149, 249
god trick 102, 141
GoGoGozo Project 53
Google Earth 94, 96–98, 100, 102, 103, 104–106, 125, 126, 129, 139, 141, 145–146, 148
Google Maps 3–4, 10, 11, 13, 126, 163–164
Google Street View 11, 13, 15, 16, 127–128, 138–153, 163–164
GPS
   device 31, 32, 35, 38, 41–42, 143–144
   satellites 143
   selective availability 139, 146, 150
   temporality 143–144
   traces 30–31, 138–153

habitus 167
hackability 250
Haiyan/Yolanda Swipe Map 199
Hanchard, M. 17, 155–172
Harley, J. B. 156, 225
Harman, G. 224
Harness, H. D. 178
Harvey, D. 124
hedonic tone 60
Henner, M. 146–148
Hind, S, 1–23, 256–267
holiday-making practices 165–168
*House-Madrid* 122
Humanitarian OpenStreetMap Team 210–219
Husserl, E. 6
hybridity 262
hyper media 96

infrastructure 161
inhabitable map 141
immutable mobile 91, 200–201
(in)formalising 17–19, 173–254, 262–265
instability 262
instantaneity 142
interfaces 161, 197–199, 206–207, 211–214

Kistler, F. 93–94
Kurgan, L. 146–147, 150

Lammes, S. 1–23, 15, 50–90, 256–267
*Landing Home in Geneva* 144, 148
latency 243, 252
Latour, B. 28, 91–92, 101, 102, 103, 199
layering 260
Lefebvre, H. 8, 159
life as contingency paradigm 231
Livehoods Project 12, 17
liveness 12–13, 243, 250
locative media 138–153
logic of inversion 92
logistics 263

MacDonald, G. 16–17, 138–153
Mackenzie, A. 142–143
mapping party 31–40
mapping projects 208–219
  as bubble 204

maps
  and territory 100–102
  in movies 96–98
  media theory 181
  relationship to photographs 141, 144–150
  rethinking 157
mashup 34, 197
Massey, D. 1, 92, 231, 257, 259
Massumi, B. 232–233
materiality 225–227, 263
McLean, K. 15, 50–90
mediality 92–94
memories 6–17, 59–62, 66–67, 106–107, 113–172, 115, 125, 128, 129–130, 260–262
Merriman, P. 1, 10, 19, 190
metagenome 223, 227, 230, 232
methodologies
  archival 209
  smell mapping 79–87
mimesis 100–102
Minard, C. J. 12, 179–180
mise-en-abyme 140
mobile media 157, 160, 169
mobility 14–15, 25–112, 82–83, 258–260
modalities of time 7, 247–248
*Mogi* 239
moments 146, 148
*Monochrome Landscapes* 146–147
mutability 3–5, 146, 200–204, 234

navigation 99–100, 100–103
navigational regime 99–100
network analysis over time 212–214
new managerialism 242
*Nine Eyes of Google Street View* 127–128, 148
*No Man's Land* 148
nodes 27–49
non-representational theory 27, 226–227

object-oriented approaches 225–227, 264
olfactory perception 53–55
ontogenetic cartography 38, 201–202, 226–227
OpenStreetMap 27–47, 210, 211, 212

Ordnance Survey Open Data 37
organising time 164–165

paramap 200
participatory model of geography 29
passenger navigation screen 33–34
*PathoMap* 18, 223–237
   as monitoring device 224
pendulum clock 143
performative mapping 61, 101
Perkins, C. 1–23, 15, 50–90, 256–267
petrichor 85
phenomenology 6
photography
   digital 115–136
   non-human 138
play 201
point of interest 42
Polak, E. 139–142
*Portraits Series* 118–119
post digital 4
practice theory 17, 159–162
   and digital maps 162–169
predictive analytics 244
preparedness 228–229, 231
present future 247
programming protocols 208
Proust, M. 115–116, 123–124

qualculation 242

Rafman, J. 127–128, 148
real time 216, 228, 229, 232, 238–251
   political implications 245
realist epistemology 242
recombination 200
relations 27–49, 203
remediation 10, 95, 96, 97, 154, 156, 243
*Res Publica* 130–133
research agendas 258–265
resistance 261
rheologists 207
rheology 197–222
rhythm 8, 28, 39, 41, 42, 43, 69–80, 92–93, 166
Robinson, A. 155
route maps 176–177, 181

Salla, C. 105
satellite imagery 96, 97, 105, 141, 145–146, 202, 212–214, 216, 219
   composite 145–146
   tiles 145
scale 103, 135, 136, 258–259
Schatzski, T. 159
*Seascape Series* (Weileder) 116–117
*Seascapes* (Sugimoto) 117–118
semiotic approaches 145
sequence 7, 246–247
site-specific sculpture 132–134
*Skydeck* 129–130
slippy map 3, 198
Sloterdijk, P. 18, 204, 206, 219
slow mapping 28, 41, 239
smell mapping 50–90
smell notes 69, 72–73
smell walk 54–58, 61, 63, 67, 69–72, 80–83
smells
   background 75
   catching 55, 83
   categorisation 73–74
   culture 62–65
   duration 59
   educative practice 62–67
   episodic 75
   history 75–76
   hunting 83
   intensity 59
   memories 59–62
   perception 54, 61
*Smellcolour sketch: Brooklyn* 56
*Smellmap: Edinburgh* 51
*Smellmap: Pamplona* 57–58, 71
*Smellmap: Paris* 66
*Smellmap: Newport, Rhode Island* 65
*Smellscaper app* 67–69
smellscapes 50–90
Snow, J. 178
social constructivist approach 225
social synchronisation 244
*Sommet* 104
space-crossed time 115–136
space-time envelopes 46
spatial cognition 10

spatial turn 8–10
spatiality 261
spatio-temporal lag 41–42
speed 191, 238, 245
Stiegler, B. 142, 187
stitching 16–17, 113–172, 260–262
strategy of deflation 91–92
Sugimoto, H. 117–119
Sutherland, T. 17–18, 175–196
syndromic surveillance 228

*Tabula Peutingeriania* 18, 20, 176–177
technicity 142
tempo 7, 245
temporal stories 57–58
temporal turn 8–13
temporality 5–6, 242–243
*Terme (le)* 129–131
thick description 159
thought figure 204, 206, 219
Thrift, N. 28, 139, 141, 175
tiles 145
time frame 5, 15, 242–243
time space compression 10, 124
time space conflation 117, 122
timelag 215
timelessness 138, 141–144, 148, 150
timing 6–7, 244–245
Tinder 12
Tomnad Project 216
transductions 142–143
translation rules 189
Transverse Mercator projection 47
Turk, C. 18, 197–222

Uber 12
urban dashboards 19, 239–246, 248–249, 250, 251
  data 240–241
  platforms 240
  politics 249–250
  purpose 241–242
  representations 248–249
  typology 240–242
  uses 240–241
  visualisation 241
urban infomatics 227–231
user centered design 249
user experience 157

van Bekkum I. 139
Van Sant, T. 145
vernacular mapping practice 27, 29
Virilio, P. 7, 121, 124–125, 135, 176, 183, 191
volatility 50–90
Volontaires internationaux en soutien aux opérations virtuelles 214, 215

ways 27–49
wearable technologies 12
Weileder, W. 16, 115–137
Wells, R. 16, 115–137
Wilmott, C. 1–23, 256–267
writing process 256

*You are Here: Information Drift* 150

zoomscape 99

Lightning Source UK Ltd.
Milton Keynes UK
UKHW011541150819
347946UK00005B/302/P